REPROGRAMMABLE CITY,
SEOUL

TERRA INCOGNITA

SNU / UPenn Collaboratives: Fall 2023

Seoul National University
Department of Architecture and Architectural Engineering
Yerin Kang Studio, Terra Incognita

University of Pennsylvania Weitzman
School of Design
David Eugin Moon Studio, The Reprogrammable City, Seoul

Sponsored by Heerim Architects & Planners

Reprogrammable City, Seoul –
Terra Incognita

TERRA INCOGNITA

REPROGRAMMABLE CITY, SEOUL

In Seoul, there exists land that has continuously existed in time but has been absent in space. This is land that does not appear on maps. Primarily consisting of various military bases left by the histories of World War II and the Korean War, these places have been isolated from the city for security reasons. Google Maps visualizes this land as dense forests or construction sites, effectively blocking spatial recognition. These unmapped lands have remained isolated like archipelagos, disconnected from the surrounding urban fabric. However, with the consecutive returns of U.S. military bases, these island-like lands are gradually opening up to the city of Seoul.

In this course, we aim to devise methods to integrate these newly accessible lands, which have been isolated from the urban context, into Seoul's spatial order. The studio will explore various types of urban projects that go beyond development-centric paradigms, aiming to explore alternative architectural projects that supplement and reconfigure the existing structure of Seoul through these new lands.
Students are expected to design their own new Seoul urbanism through research. Fundamentally, concepts like 'recovering' the land, 'connecting' with the surroundings, 'relating' to the city and nature, and 'emptying' the land will take precedence over simply 'building and filling' the site.

After determining how to handle this land through their unique Seoul urbanism, students will design architecture or spaces that remember the temporality of this land. What architectural strategies will deal with memory and traces in land layered with historical contexts? The history here is not only human history. It also includes the temporality of non-human life forms that have long inhabited the land. Attention should be paid to the ecological map that has persisted outside the history of humanism.

SNU

Terra Incognita (지도에도 없는 땅)

담당교수: 강예린

서울에는 시간속에 줄곧 존재해왔으나 공간으로는 부재해온 땅이 있다. 지도에 없는 땅이다. 주로 2차 세계대전과 한국전쟁의 역사가 남긴 각종 군사기지로서, 보안을 이유로 도시와 담을 쌓은 곳이다. 구글지도는 이 땅을 울창한 나무의 숲이나 공사중인 나대지로 가시화함으로써, 공간 인지를 차단한다. 이 지도에도 없는 땅들은 주변 도시조직과 관계를 맺지 않고 마치 군도 처럼 띄엄띄엄 고립되어 왔다. 그러나 연이은 미군기지의 반환으로 이 섬과 같은 땅들이 서울 도시에 차근차근 열리고 있다.

이 수업에서 우리는 막 도시에 열린 땅, 그러나 도시의 컨텍스트와 상관없는 섬과 같은 대지를 서울의 공간질서 안에 자리잡게 만드는 방법을 고안하려 한다. 스튜디오는 개발 중심 패러다임을 넘어서는 다양한 유형의 도시 프로젝트의 가능성을 탐색하며, 새로운 땅으로 인해 기존 서울의 구조가 보완되고 재형성되는 대안의 건축 프로젝트를 탐구하는 것이 목표다.

학생들은 리서치를 통해 본인의 새로운 서울 어바니즘을 고안해야 한다. 기본적으로 대지를 '회복하기', 주변과 '연결하기', 도시 및 자연과 '관계 맺기', 땅을 비우기 같은 개념이 필지 안에 '짓고 채우기'보다 우선시하는 방식이 될 것이다.

각자의 서울 어바니즘을 통해 이 대지를 다루는 방식을 정한 후, 이 땅의 시간성을 기억하는 건축 혹은 공간을 디자인한다. 역사적 맥락이 켜켜이 쌓인 땅에서 어떤 건축전략으로 기억과 흔적을 다룰 것인가? 여기서의 역사는 인간의 역사만이 아니다. 땅에서 오랫동안 서식해온 인간 이외의 생명체들에 대한 시간성도 포함된다. 인본주의의 역사 밖에서 지속해온 생태적인 지도를 주목해야 한다.

TERRA INCOGNITA

REPROGRAMMABLE CITY, SEOUL

As a continuation of a series of collaborative studios with PennDesign and Seoul National University, The Reprogrammable City, Seoul traveling studio challenges the notion of "destruction for progress" as the default and unquestioned solution for renewal and aims to provoke alternative paradigms for transformation and change. Engaging a selection of sites that each deals with conditions of "Post-____" urbanisms, where the existing environments become novel sites that provoke new architectural strategies and potentials, the projects will engage in some of the most urgent challenges in the contemporary environment, while posting new projects of the "afterwards."

Korea has also been influenced by a long and entangled history manifested with global-scale social and political events, more recently marked by the Korean War, its colonialized era under Japanese rule which concluded at the end of World War II, and the proxied and continuing tension between the North and South. Modern Korea and Seoul's exponential growth since the end of the Korean War engaged the spatial apparatus of development aggressively, only to its inevitable perfection. Not only did the city structure itself with robust monuments to the ideals of production, exchange, and movement, it successfully established a system of perpetual growth through its housing supply mechanism, and construction projects. Seoul has been one immense site of continual "improvement" and accumulation, imploding from its dizzying density and vibrant productivity, producing 23 similarly dense satellite cities. Covering an area relative to the size of metropolitan Detroit but 14 times its population, Seoul seemed to continue in a trajectory of unstoppable growth, until recent confrontations with the realities of a city that is, in many regards, already "developed."

As the City of Seoul administration had attempted to reduce "redevelopment" zones in the city, and redirect its attention to the possibilities in a different set of urban project typologies that "goes beyond the development-centric paradigm", the studio aims to explore the potentials of alternative kinds of architectural projects to reshape the city. Seoul, one of the first "developed" cities in Asia, is a testing ground for redefining the future contemporary city, an opportunity to consider notions of the different publics and their needs, while carefully challenging accepted and ubiquitously applied familiar rules of amelioration through the (re)invention of typologies that reintroduces the civic to the urban.

The studio will travel to Korea to observe and document the site and its contexts, and conversations with practitioners and other locals will also inform the investigation. Working in teams, students will work from a selection of sites and programs that will explore engaging the context of Seoul and its layered past, and may suggest other sites with approval from the instructor.

UPENN

The Reprogrammable City, Seoul

재프로그래밍 가능한 도시, 서울

지도 교수: David Eugin Moon

PennDesign과 서울대학교와의 일련의 협업 스튜디오인 The Reprogrammable City의 연속인 서울 여행 스튜디오는 "진보를 위한 파괴"라는 개념에 도전하고 갱신을 위한 의심할 여지 없는 해결책을 제시하며 변화와 변화를 위한 대안적 패러다임을 불러일으키는 것을 목표로 합니다. 기존 환경이 새로운 건축 전략과 잠재력을 불러일으키는 새로운 장소가 되는 "포스트-____" 도시화의 조건을 다루는 다양한 장소를 참여시키는 프로젝트는 현대 환경에서 가장 시급한 과제 중 일부에 참여할 것입니다. "그 후"의 새로운 프로젝트를 게시합니다.

한국은 또한 세계적 규모의 사회적, 정치적 사건으로 드러나는 길고 얽힌 역사의 영향을 받아 왔으며, 최근에는 한국전쟁, 즉 제2차 세계대전 말기로 끝난 일본의 식민지 시대를 겪었습니다. 최근에는 한국전쟁, 제2차 세계대전으로 종결된 일제의 식민지 시대, 남북 간 대리적이고 지속되는 긴장 등 세계적 규모의 사회적, 정치적 사건으로 드러난 길고 얽힌 역사의 영향을 받았습니다. 그리고 남쪽. 한국전쟁 이후 현대 한국과 서울의 기하급수적인 성장은 공간적 발전 장치를 적극적으로 활용했지만 필연적으로 완벽해졌습니다. 도시 구조 자체가 생산, 교환, 유통의 이상을 상징하는 견고한 기념물을 갖추고 있었을 뿐만 아니라 이후 주택공급 메커니즘과 건설사업을 통해 영속성장 체제를 성공적으로 구축했습니다. 서울은 지속적인 '개선'과 축적의 거대한 현장 중 하나였으며, 현기증나는 밀도와 활기찬 생산성으로 인해 비슷한 밀도의 위성 도시가 23개나 탄생했습니다. 대도시 디트로이트 크기에 비해 면적이 넓지만 인구는 14배에 달하는 서울은 최근 여러 측면에서 이미 '개발'된 도시의 현실과 직면하기 전까지는 멈출 수 없는 성장의 궤적을 계속 이어가는 것처럼 보였습니다.

서울시 행정부가 도시의 "재개발" 구역을 줄이고 "개발 중심 패러다임을 넘어서는" 다양한 도시 프로젝트 유형의 가능성에 관심을 돌리려고 시도함에 따라 스튜디오는 도시를 재형성하기 위한 대안적인 종류의 건축 프로젝트의 잠재력. 아시아 최초의 "개발된" 도시 중 하나인 서울은 미래의 현대 도시를 재정의하기 위한 시험장이며, 다양한 대중의 개념과 그들의 요구를 고려할 수 있는 기회이자, 수용되고 어디에나 적용되는 친숙한 개선 규칙에 신중하게 도전하는 기회입니다. 시민을 도시에 다시 도입하는 유형학의 (재)발명.

스튜디오는 현장과 그 맥락을 관찰하고 기록하기 위해 한국을 방문할 예정이며, 실무자 및 다른 현지인들과의 대화도 조사에 대한 정보를 제공할 것입니다. 팀으로 활동하면서 학생들은 서울의 맥락과 다층적인 과거를 탐구하는 다양한 장소와 프로그램을 통해 작업하게 되며 강사의 승인을 받아 다른 장소를 제안할 수도 있습니다.

REPROGRAMMABLE CITY, SEOUL

TERRA INCOGNITA

CONTENTS

I
YONGSAN

II
SONG HYEON DONG

III
FAR EAST BASE

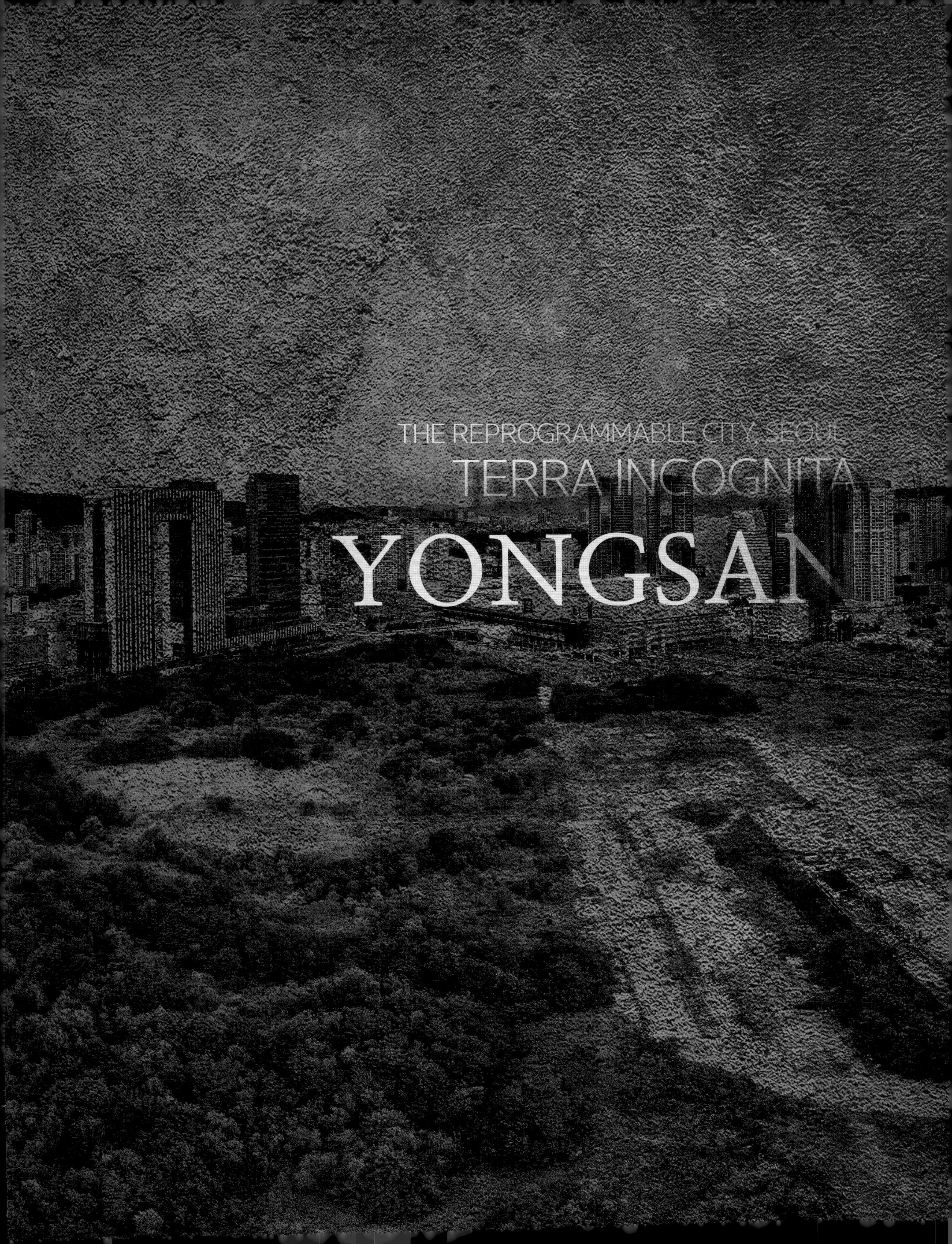

THE REPROGRAMMABLE CITY, SEOUL

TERRA INCOGNITA

YONGSAN

TERRA INCOGNITA

I
YONGSAN

01
SNU
1-1) TEAM SITE SURVEY ANALYSIS
1-2) STUDENTS' WORKS
- Park, Young Jo
- Lee, Tae Yang
- Cho, Dong Hyun

02
UPENN
1-1) TEAM SITE SURVEY ANALYSIS
1-2) STUDENTS' WORKS
- Diego Martin + Hei Wai Valerie Tse
- Maria Jose + Kelvin Yu

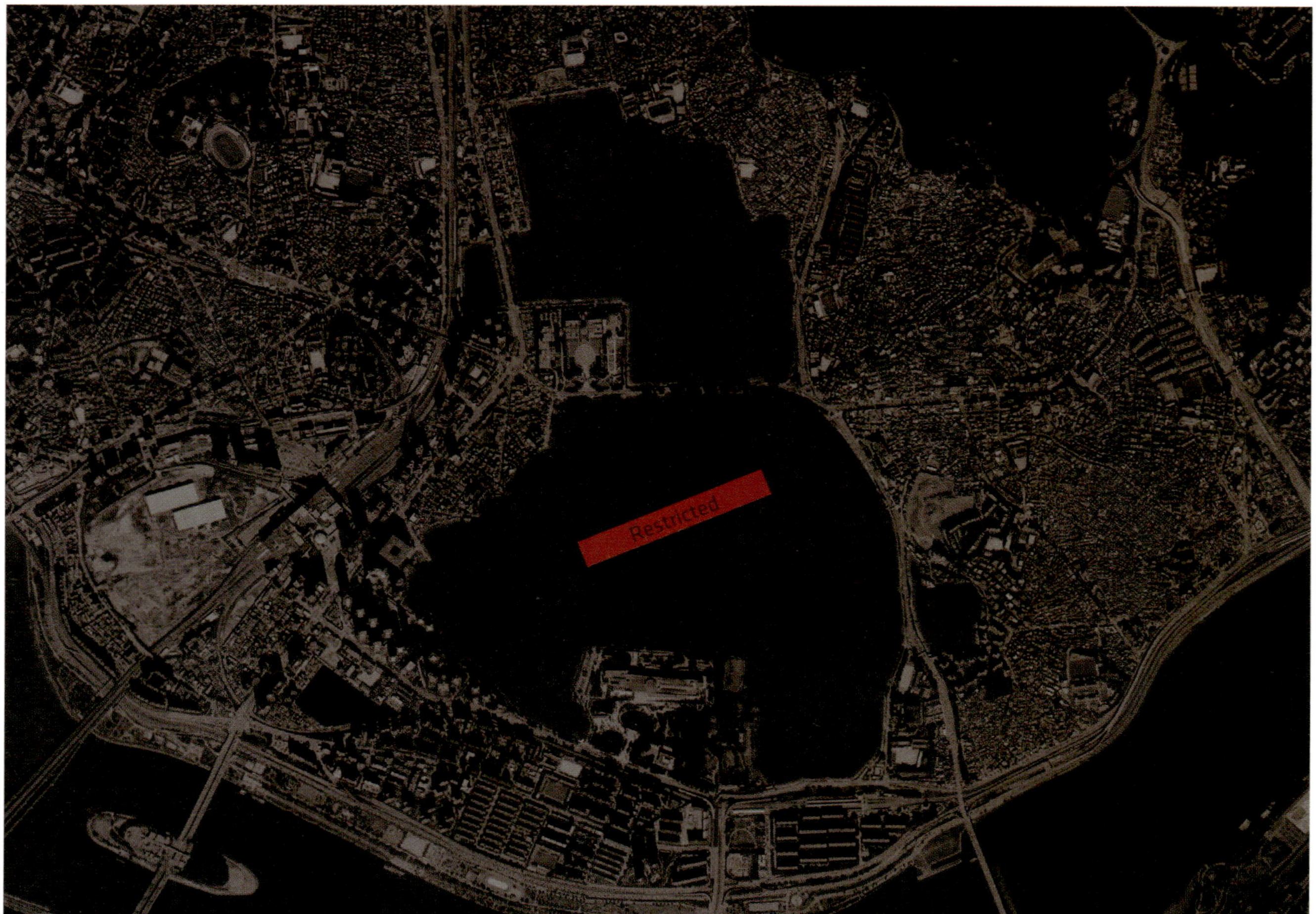

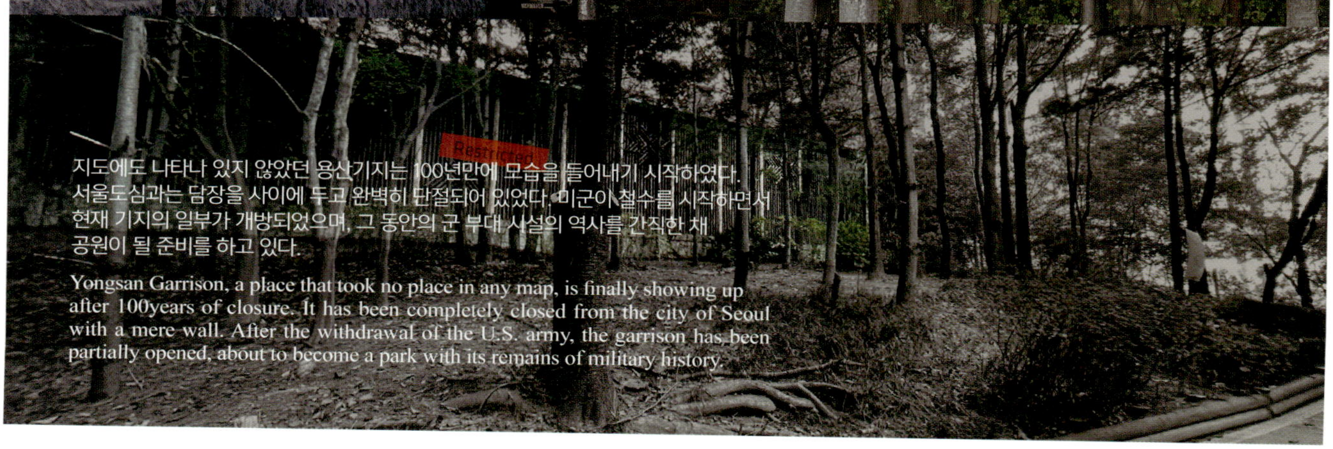

지도에도 나타나 있지 않았던 용산기지는 100년만에 모습을 들어내기 시작하였다. 서울도심과는 담장을 사이에 두고 완벽히 단절되어 있었다. 미군이 철수를 시작하면서 현재 기지의 일부가 개방되었으며, 그 동안의 군 부대 시설의 역사를 간직한 채 공원이 될 준비를 하고 있다.

Yongsan Garrison, a place that took no place in any map, is finally showing up after 100years of closure. It has been completely closed from the city of Seoul with a mere wall. After the withdrawal of the U.S. army, the garrison has been partially opened, about to become a park with its remains of military history.

Resident Parking
Only
7073 B

용산기지는 점진적 개방이 결정된 이후 도시개발, 정부시설, 역사적 관성, 생태적 복원 네 가지 큰 힘들의 영향을 받고 있다.
점선으로 표시한 대상지가 네 가지의 영향력과 가장 밀접하게 닿아 있는 곳이라고 생각했다.

Yongsan Garrison, with the decision of opening the place to the public, is being influenced by four different forces; capitalism, government, history, and ecology.
We suggest that the land marked with dotted line is the most vulnerable place to those influences.

GOVERNMENT
ECOLOGY
N
0 20 50 100 200

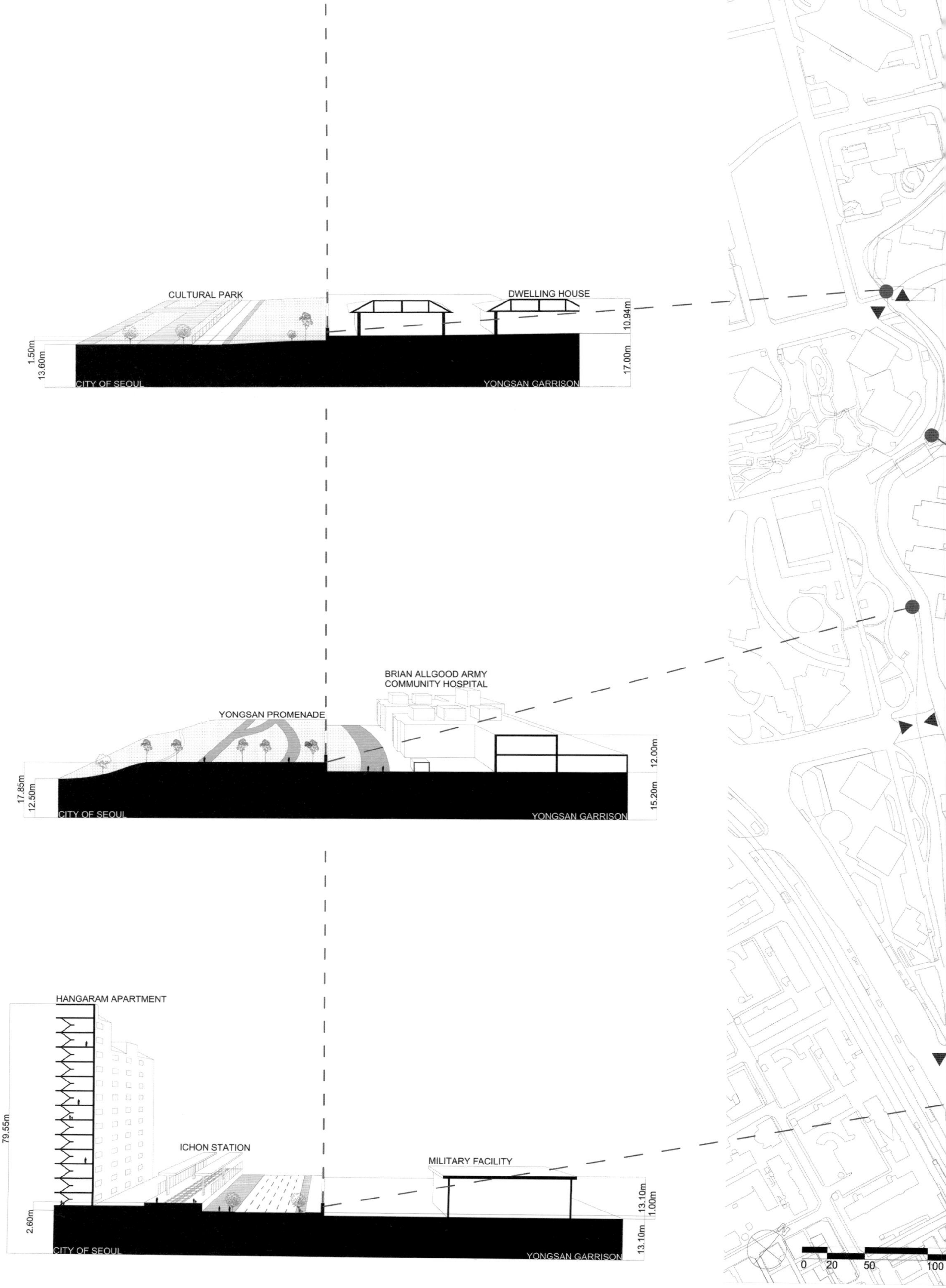
CULTURAL PARK
DWELLING HOUSE
1.50m
13.60m
10.94m
17.00m
CITY OF SEOUL
YONGSAN GARRISON
BRIAN ALLGOOD ARMY
COMMUNITY HOSPITAL
YONGSAN PROMENADE
17.85m
12.50m
12.00m
15.20m
CITY OF SEOUL
YONGSAN GARRISON
HANGARAM APARTMENT
79.55m
ICHON STATION
MILITARY FACILITY
2.60m
13.10m
1.00m
13.10m
CITY OF SEOUL
YONGSAN GARRISON
0
20
50
100

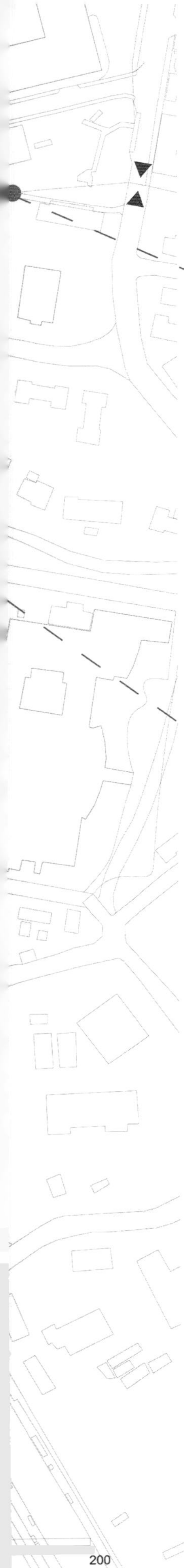

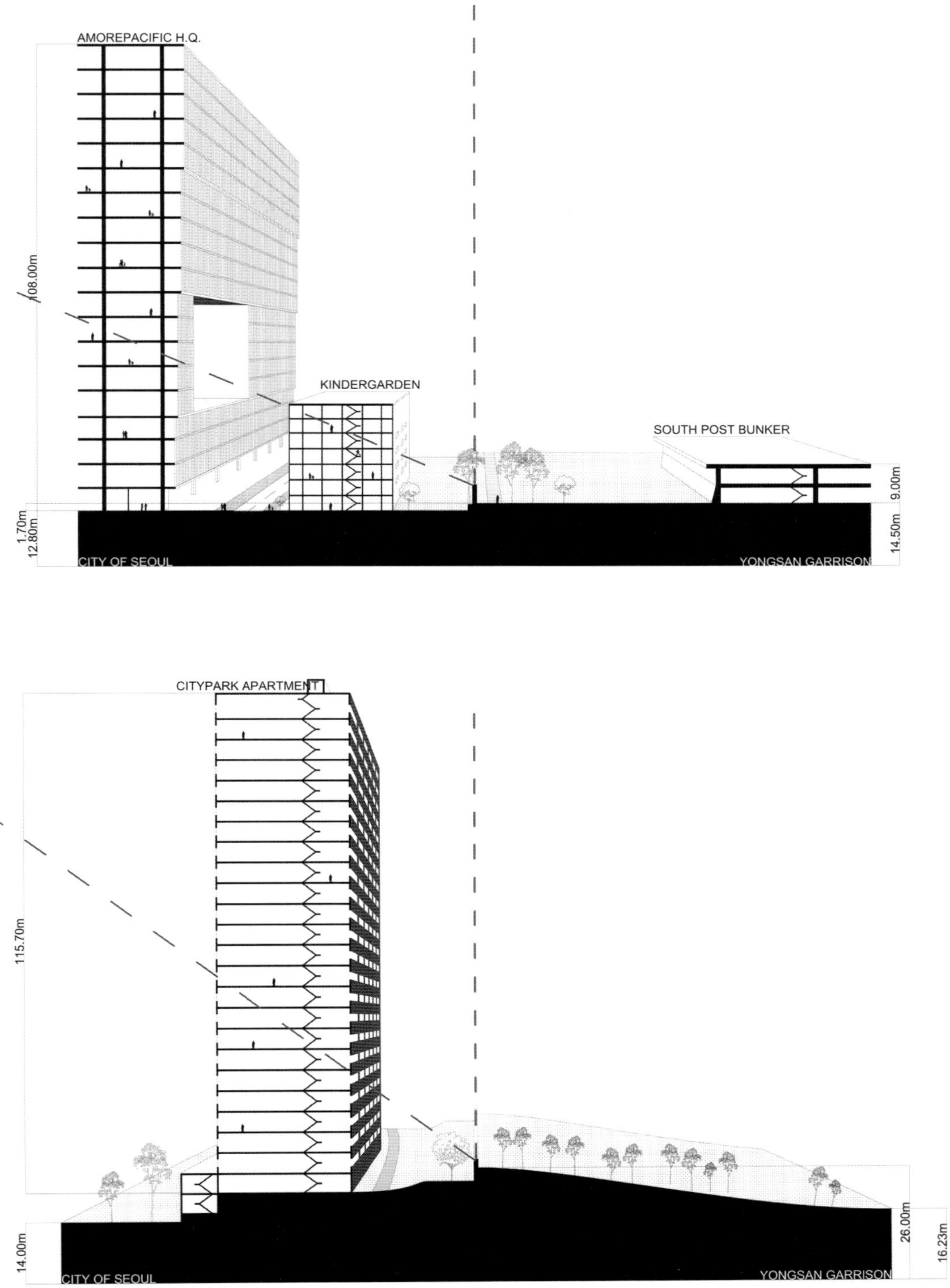

용산기지와 도시는 땅의 높이 차이가 존재하며 그 경계에서
다양한 성격을 지닌 서로 다른 땅들이 마주하고 있다.
경계를 사이에 두고 도시에는 기업 본사, 아파트 단지, 공원 등이 있고,
용산기지에는 벙커, 숙소, 녹지, 병원 등이 있으며, 철조망 뿐만 아니라
변화하는 땅의 높이 차이도 경계를 형성한다.

There is ground level difference on the boundary between Yongsan Garrison and City of Seoul, where different lands with many characters meet. On the city side there are a corporate H.Q., apartments, a park and etc while the garrison consists of a bunker, a hospital, small houses for soldiers, grass and so on. Not only wired walls but also the ground level difference creates a boundary between them.

Height of building around site
Site
~25m
25m~50m
50m~100m
100m~

1969 aerial photograph, three towns around the Garrison
Haebangchon
Samgackji
Itaewon
Site

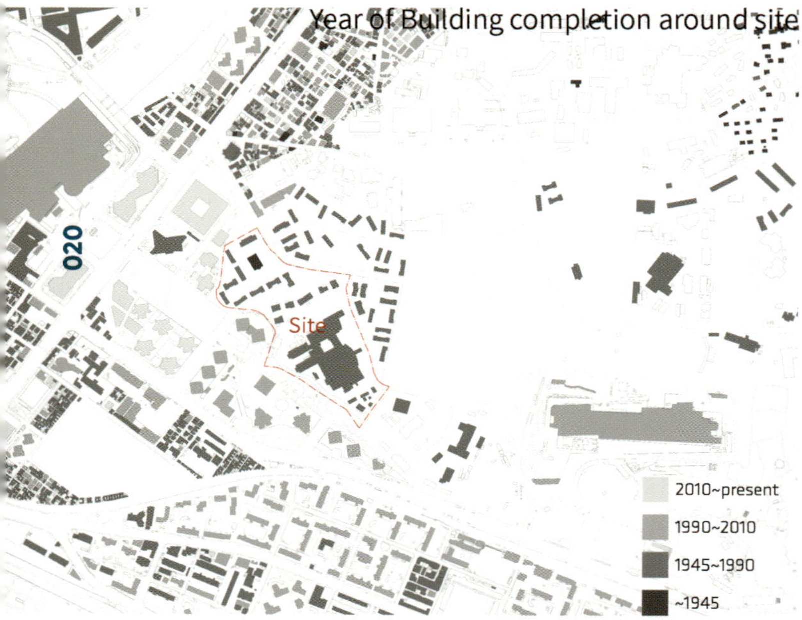
Year of Building completion around site
Site
2010~present
1990~2010
1945~1990
~1945

Bus stations with main streets
Site
Bus station

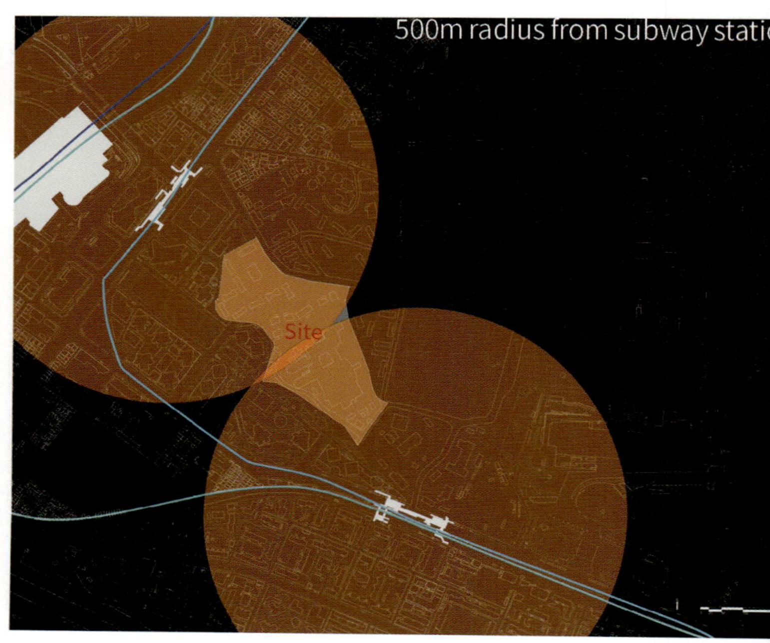
500m radius from subway static
Site

서울과 기지는 생태적, 도시적으로 같은 맥락을 갖기 위해 노력하고 있다.
기지 안팎의 상황은 밀도의 측면에서도, 교통의 측면에서도 달라왔다.
공원으로 탈바꿈하기 위한 노력의 과정을 거치면서,
1970년대 미군의 영향을 받아 생겨난 마을들도 생태의 차원에서 서로 연결되며,
도시적 맥락을 공유할 가능성을 가진다.

Seoul and the garrison are making efforts to gradually share a similar ecological and urban context. The situations within and around the bases have differed in terms of density and transportation.
As efforts are made to transform them into parks, villages that emerged under the influence of the US military in the 1970s could also be interconnected in ecological dimensions, potentially sharing urban contexts.

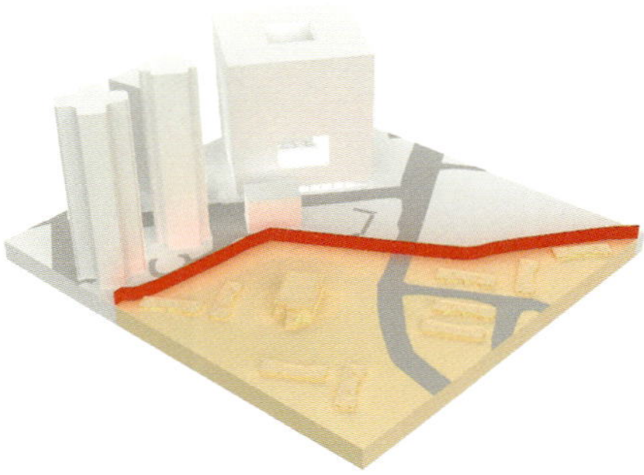
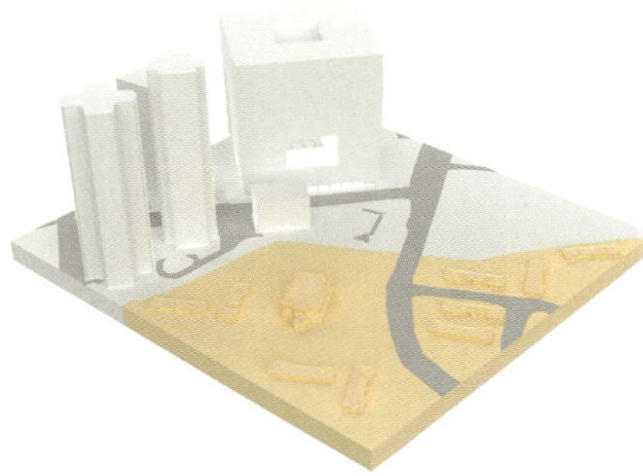
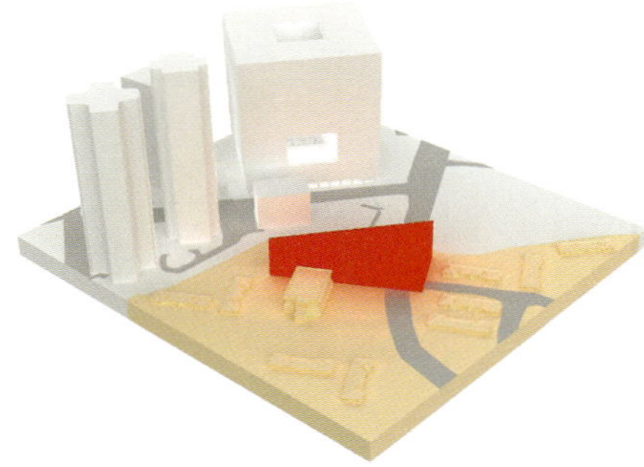
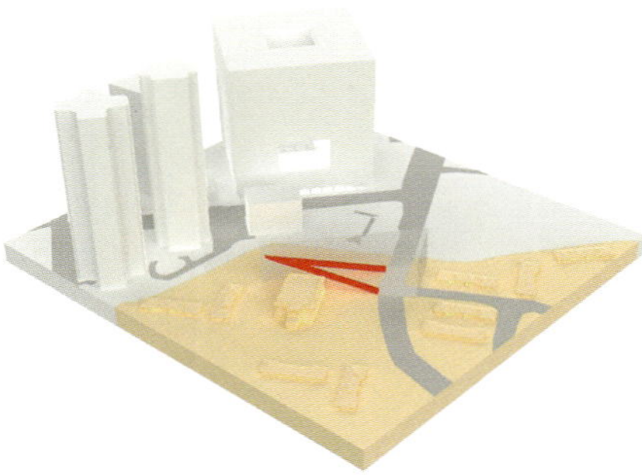
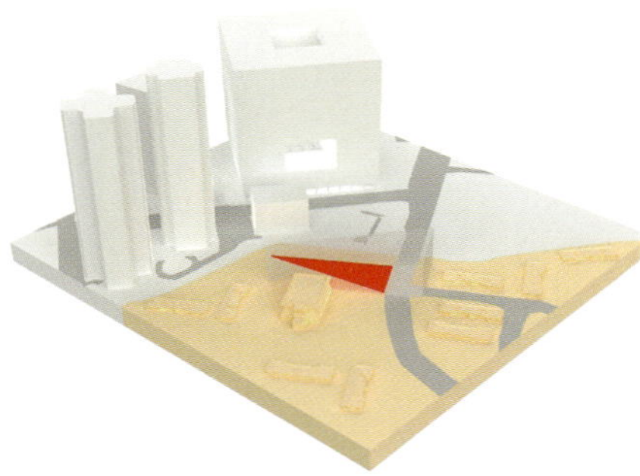
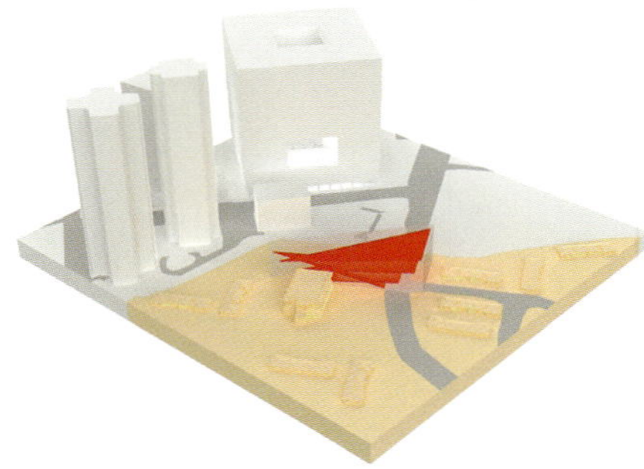
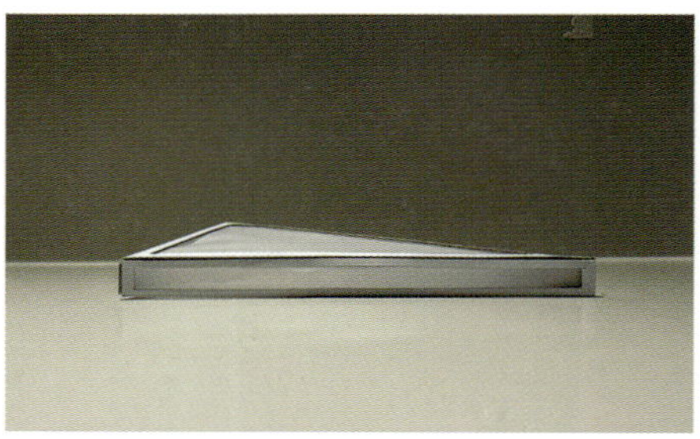
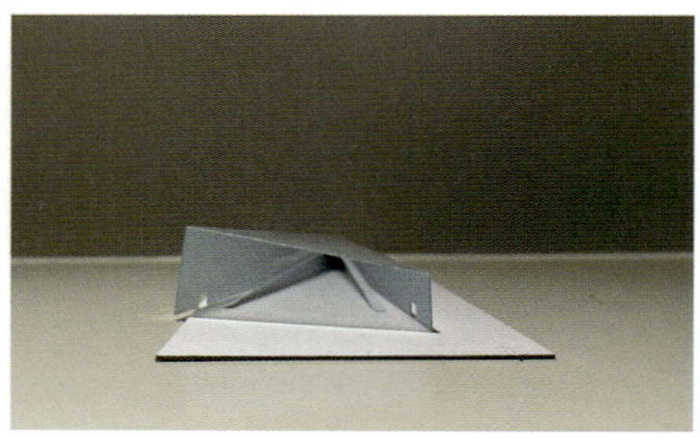
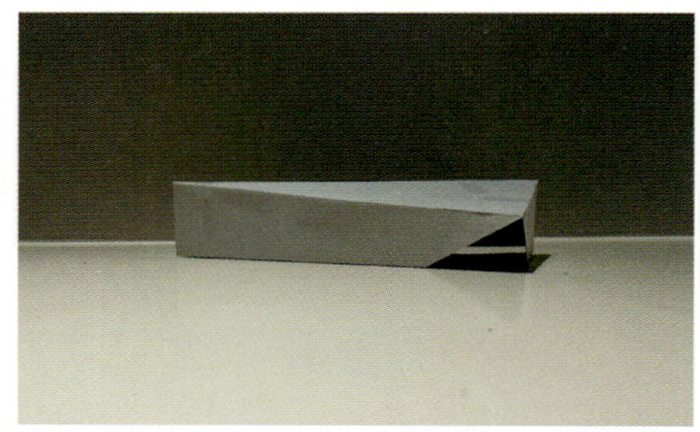
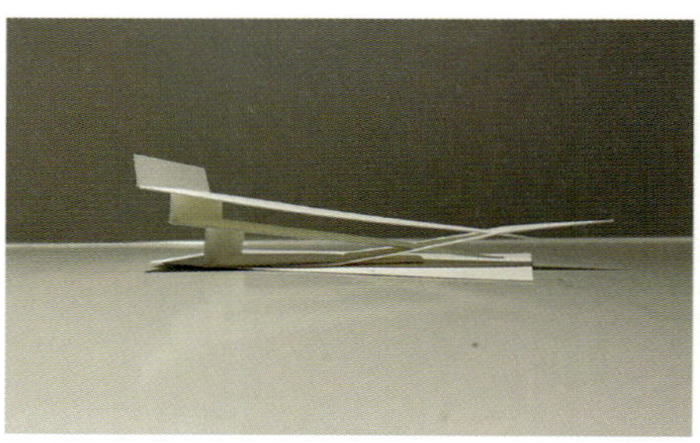

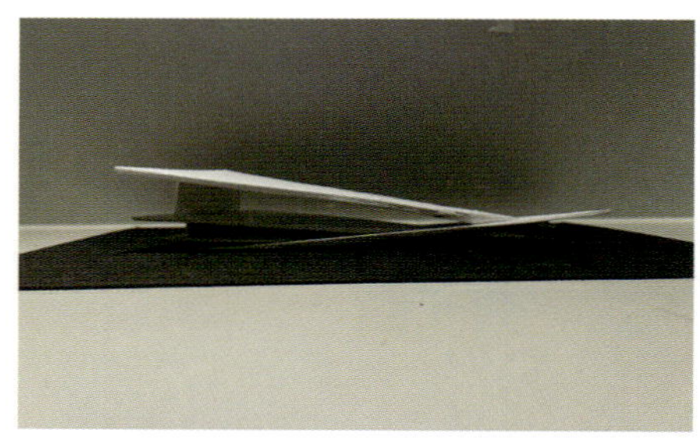
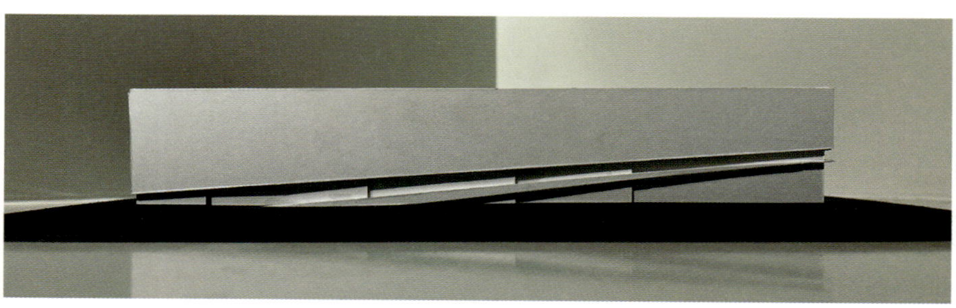

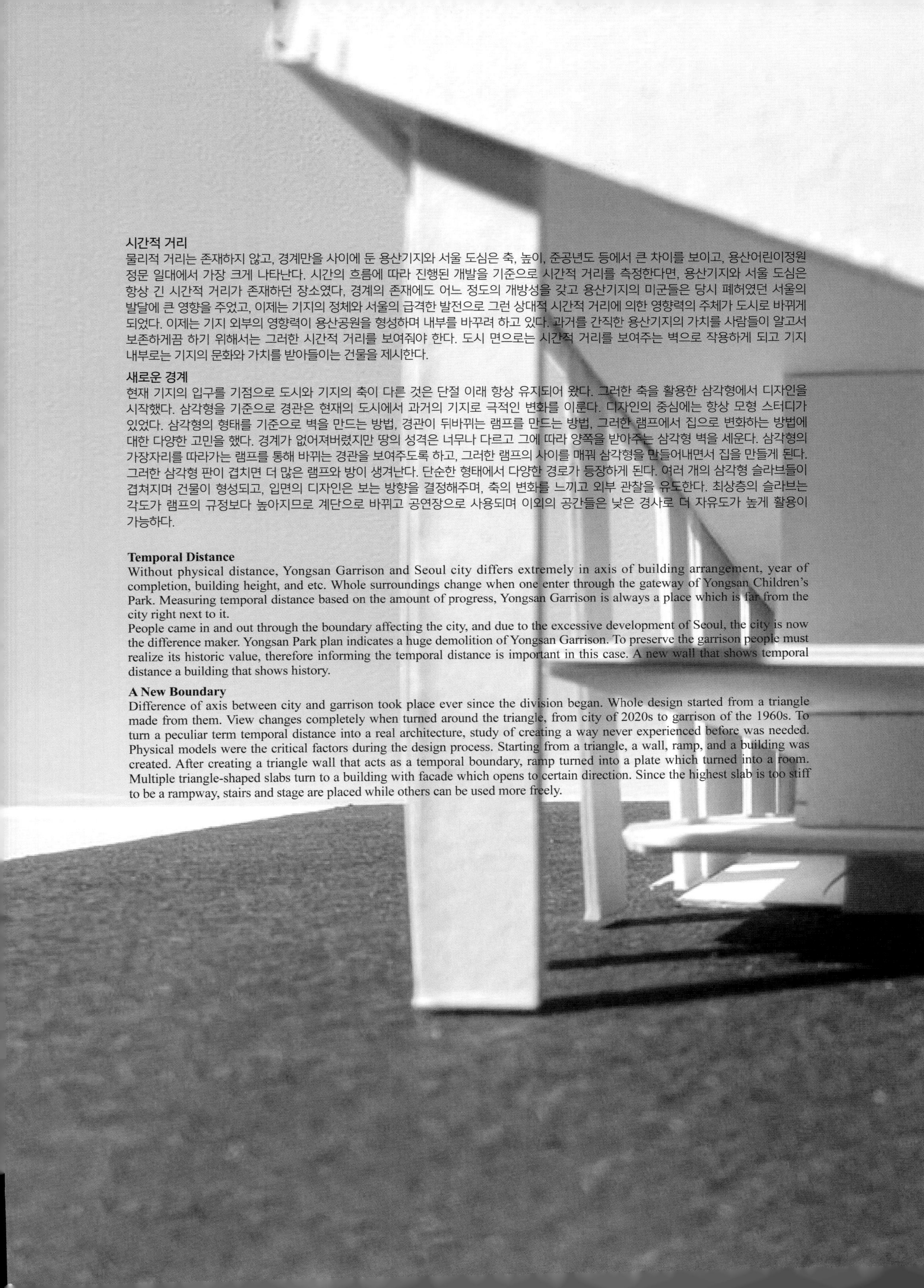

시간적 거리

물리적 거리는 존재하지 않고, 경계만을 사이에 둔 용산기지와 서울 도심은 축, 높이, 준공년도 등에서 큰 차이를 보이고, 용산어린이정원 정문 일대에서 가장 크게 나타난다. 시간의 흐름에 따라 진행된 개발을 기준으로 시간적 거리를 측정한다면, 용산기지와 서울 도심은 항상 긴 시간적 거리가 존재하던 장소였다. 경계의 존재에도 어느 정도의 개방성을 갖고 용산기지의 미군들은 당시 폐허였던 서울의 발달에 큰 영향을 주었고, 이제는 기지의 정체와 서울의 급격한 발전으로 그런 상대적 시간적 거리에 의한 영향력의 주체가 도시로 바뀌게 되었다. 이제는 기지 외부의 영향력이 용산공원을 형성하며 내부를 바꾸려 하고 있다. 과거를 간직한 용산기지의 가치를 사람들이 알고서 보존하게끔 하기 위해서는 그러한 시간적 거리를 보여줘야 한다. 도시 면으로는 시간적 거리를 보여주는 벽으로 작용하게 되고 기지 내부로는 기지의 문화와 가치를 받아들이는 건물을 제시한다.

새로운 경계

현재 기지의 입구를 기점으로 도시와 기지의 축이 다른 것은 단절 이래 항상 유지되어 왔다. 그러한 축을 활용한 삼각형에서 디자인을 시작했다. 삼각형을 기준으로 경관은 현재의 도시에서 과거의 기지로 극적인 변화를 이룬다. 디자인의 중심에는 항상 모형 스터디가 있었다. 삼각형의 형태를 기준으로 벽을 만드는 방법, 경관이 뒤바뀌는 램프를 만드는 방법, 그러한 램프에서 집으로 변화하는 방법에 대한 다양한 고민을 했다. 경계가 없어져버렸지만 땅의 성격은 너무나 다르고 그에 따라 양쪽을 받아주는 삼각형 벽을 세운다. 삼각형의 가장자리를 따라가는 램프를 통해 바뀌는 경관을 보여주도록 하고, 그러한 램프의 사이를 매꿔 삼각형을 만들어내면서 집을 만들게 된다. 그러한 삼각형 판이 겹치면 더 많은 램프와 방이 생겨난다. 단순한 형태에서 다양한 경로가 등장하게 된다. 여러 개의 삼각형 슬라브들이 겹쳐지며 건물이 형성되고, 입면의 디자인은 보는 방향을 결정해주며, 축의 변화를 느끼고 외부 관찰을 유도한다. 최상층의 슬라브는 각도가 램프의 규정보다 높아지므로 계단으로 바뀌고 공연장으로 사용되며 이외의 공간들은 낮은 경사로 더 자유도가 높게 활용이 가능하다.

Temporal Distance

Without physical distance, Yongsan Garrison and Seoul city differs extremely in axis of building arrangement, year of completion, building height, and etc. Whole surroundings change when one enter through the gateway of Yongsan Children's Park. Measuring temporal distance based on the amount of progress, Yongsan Garrison is always a place which is far from the city right next to it.

People came in and out through the boundary affecting the city, and due to the excessive development of Seoul, the city is now the difference maker. Yongsan Park plan indicates a huge demolition of Yongsan Garrison. To preserve the garrison people must realize its historic value, therefore informing the temporal distance is important in this case. A new wall that shows temporal distance a building that shows history.

A New Boundary

Difference of axis between city and garrison took place ever since the division began. Whole design started from a triangle made from them. View changes completely when turned around the triangle, from city of 2020s to garrison of the 1960s. To turn a peculiar term temporal distance into a real architecture, study of creating a way never experienced before was needed. Physical models were the critical factors during the design process. Starting from a triangle, a wall, ramp, and a building was created. After creating a triangle wall that acts as a temporal boundary, ramp turned into a plate which turned into a room. Multiple triangle-shaped slabs turn to a building with facade which opens to certain direction. Since the highest slab is too stiff to be a rampway, stairs and stage are placed while others can be used more freely.

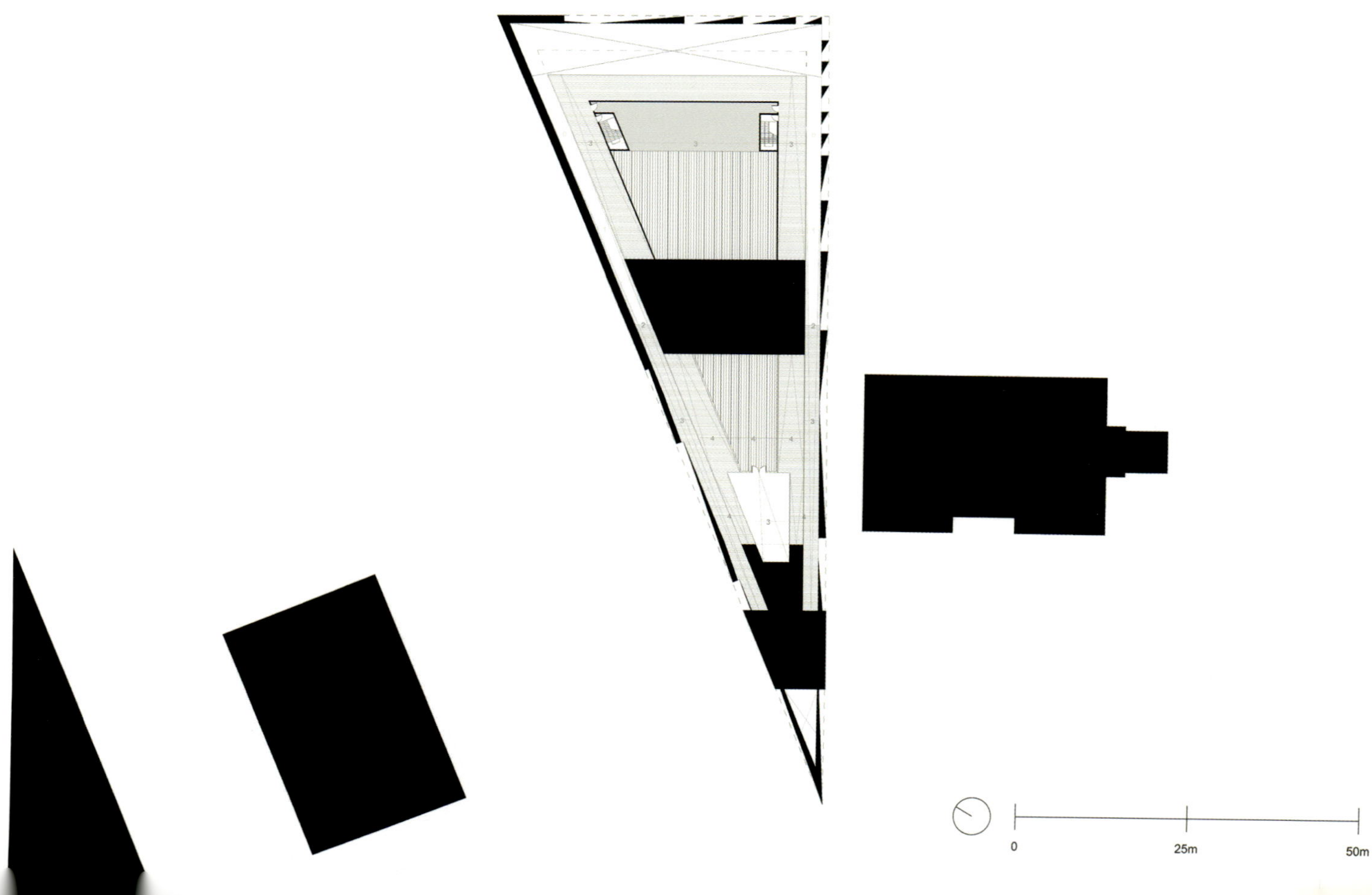
0
25m
50m

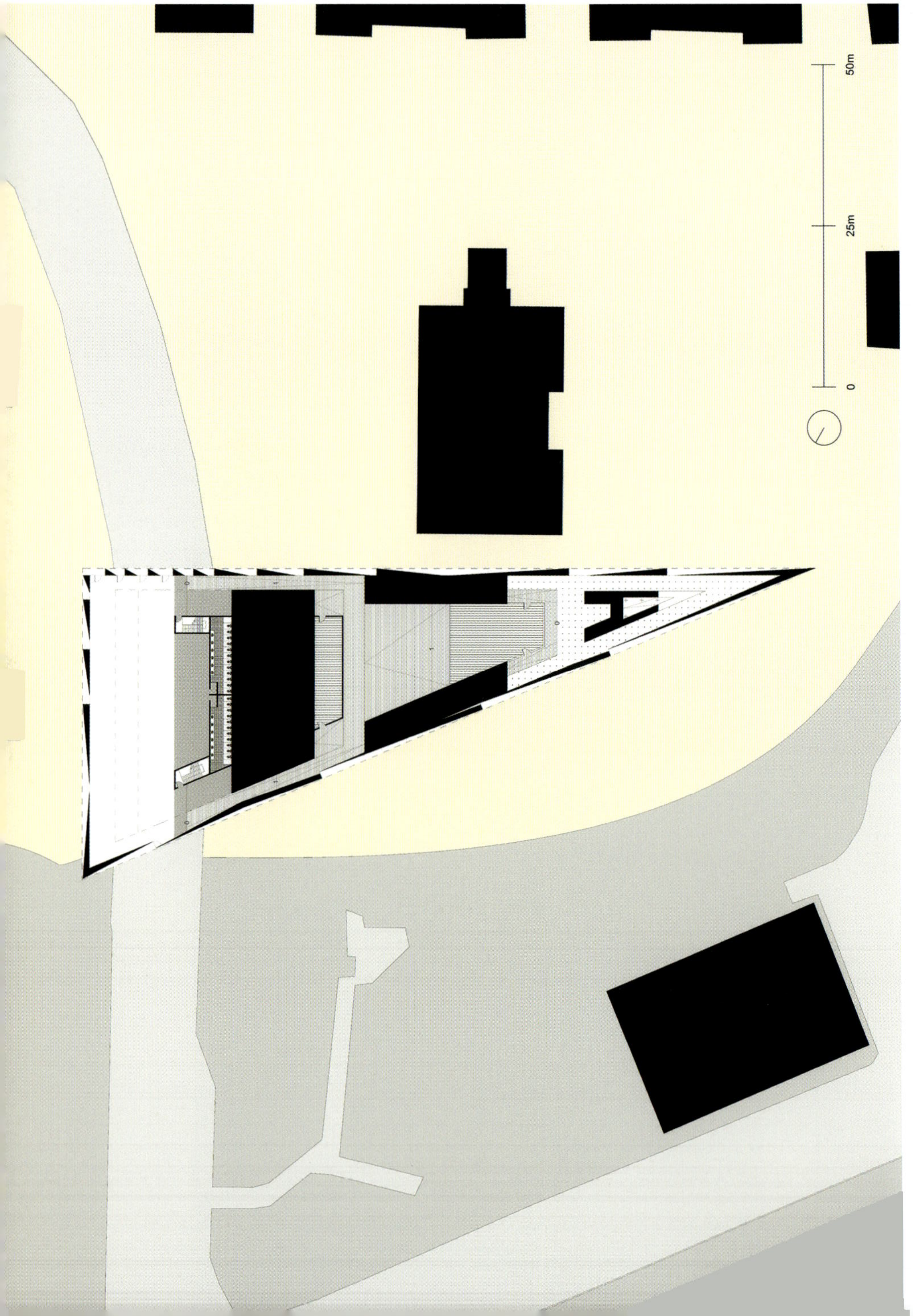
0
25m
50m

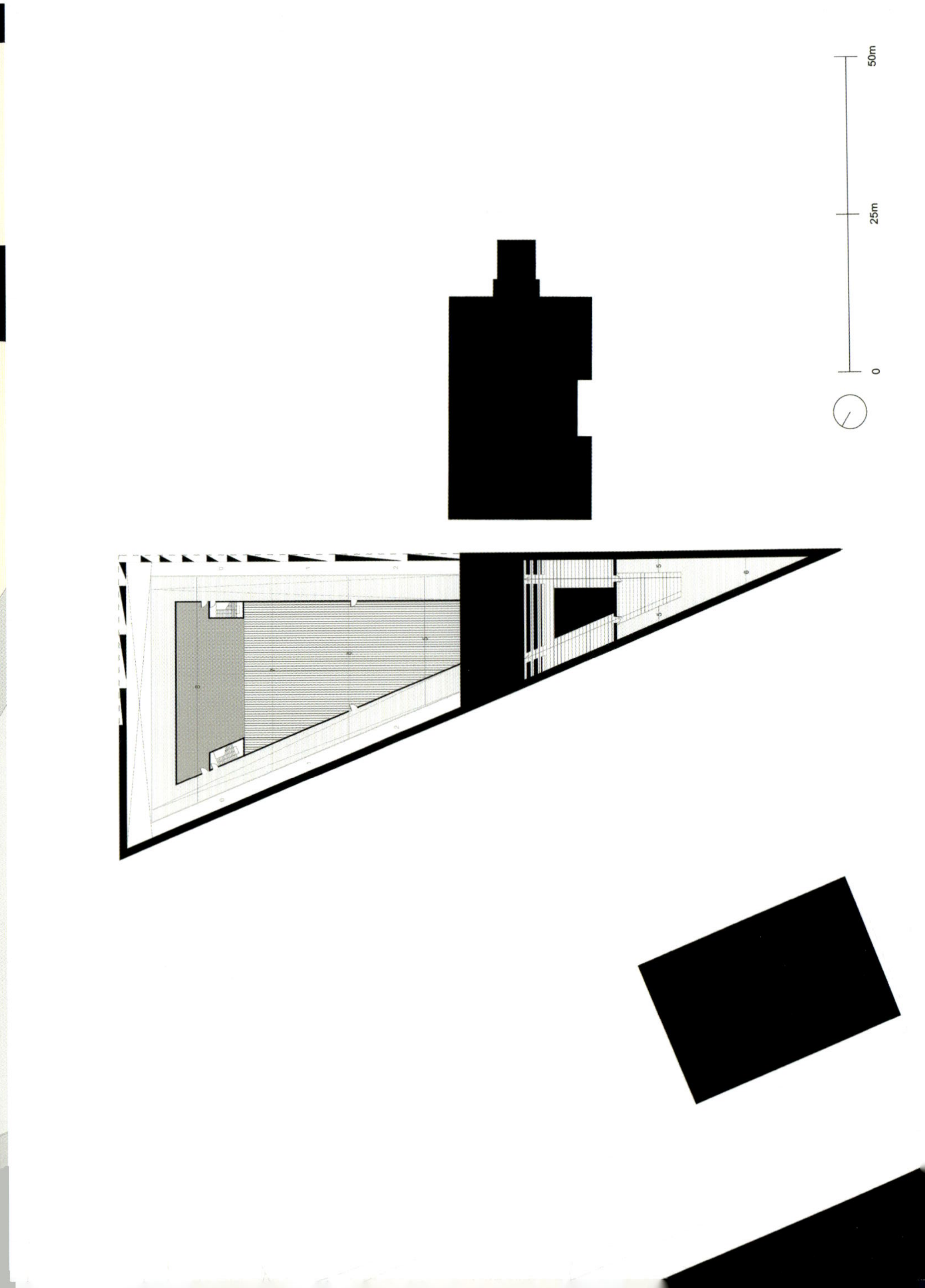
0
25m
50m

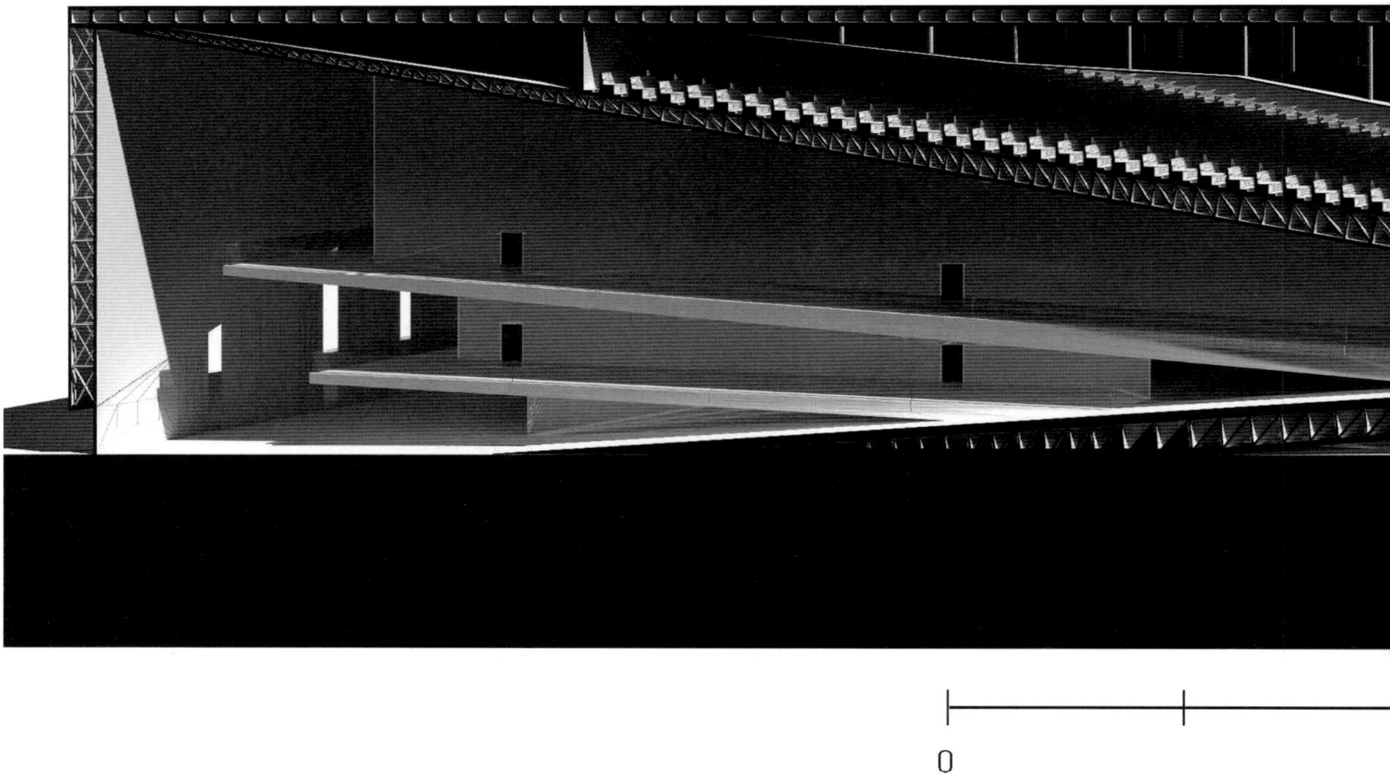
0

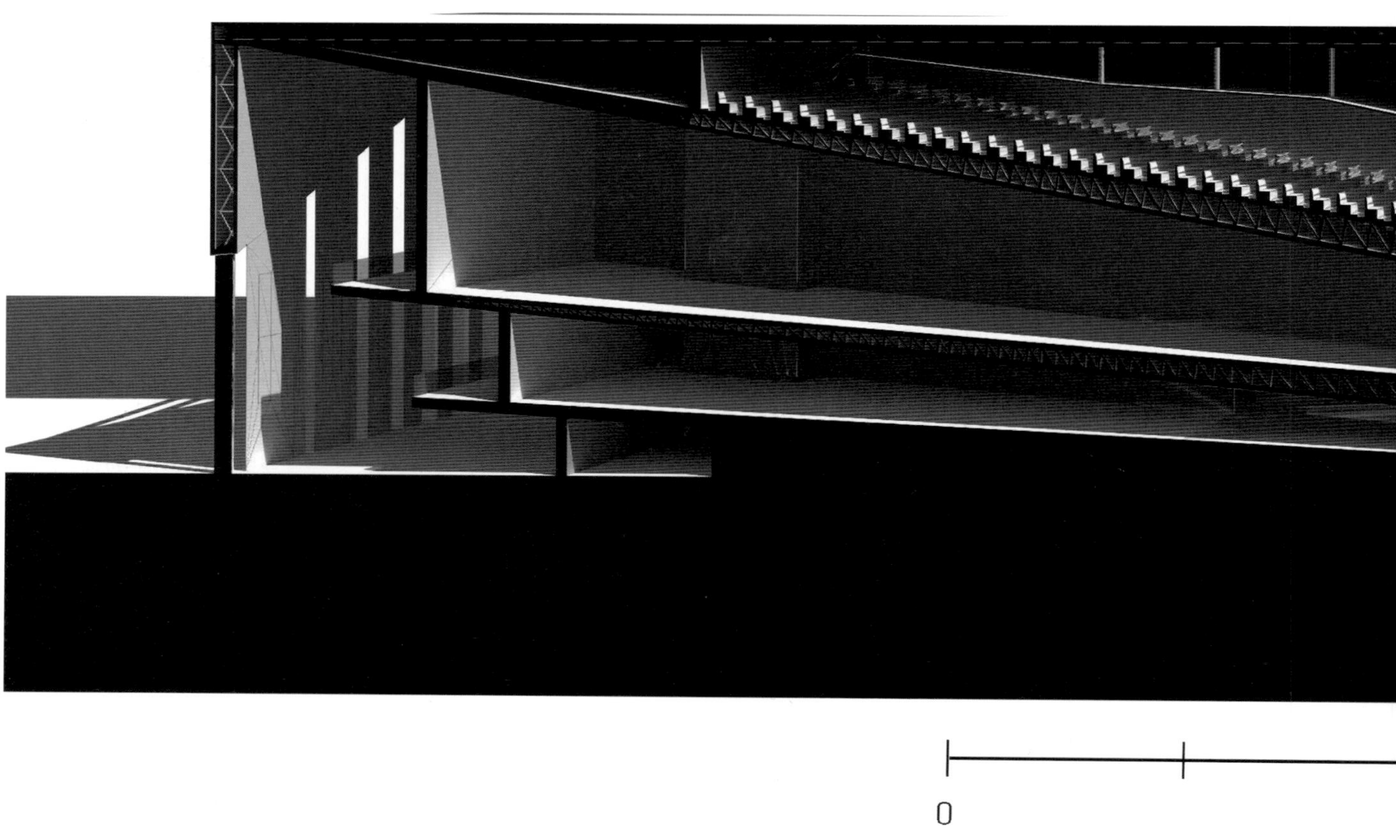
0

50m

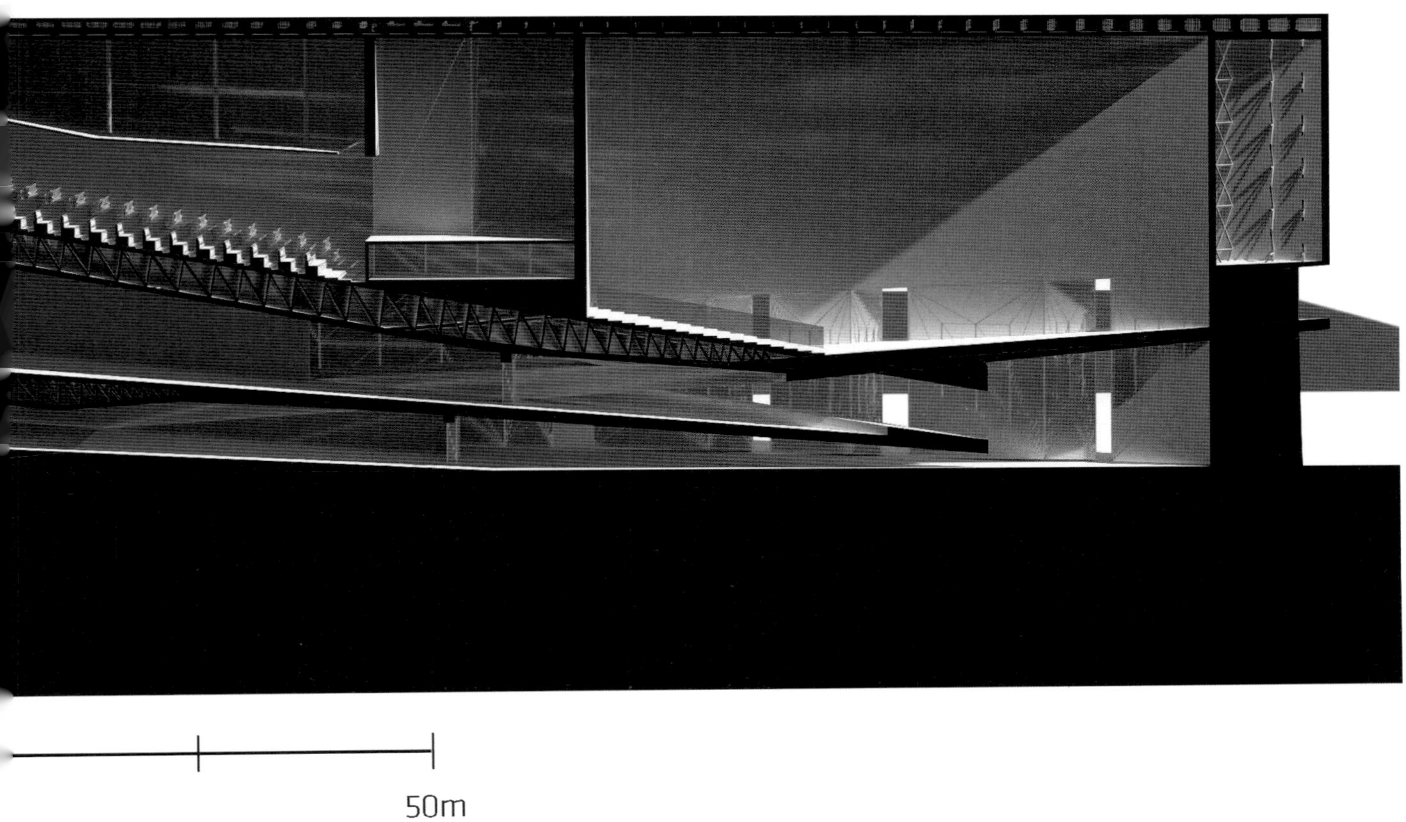
50m

Yonsan Station
Yongsan Gateway
Janpanese bunker
Children's Garden
Ministry of National Defense
U.S. HOSPITAL
Future Yongsan Park
Yongsan City
Ichon Station

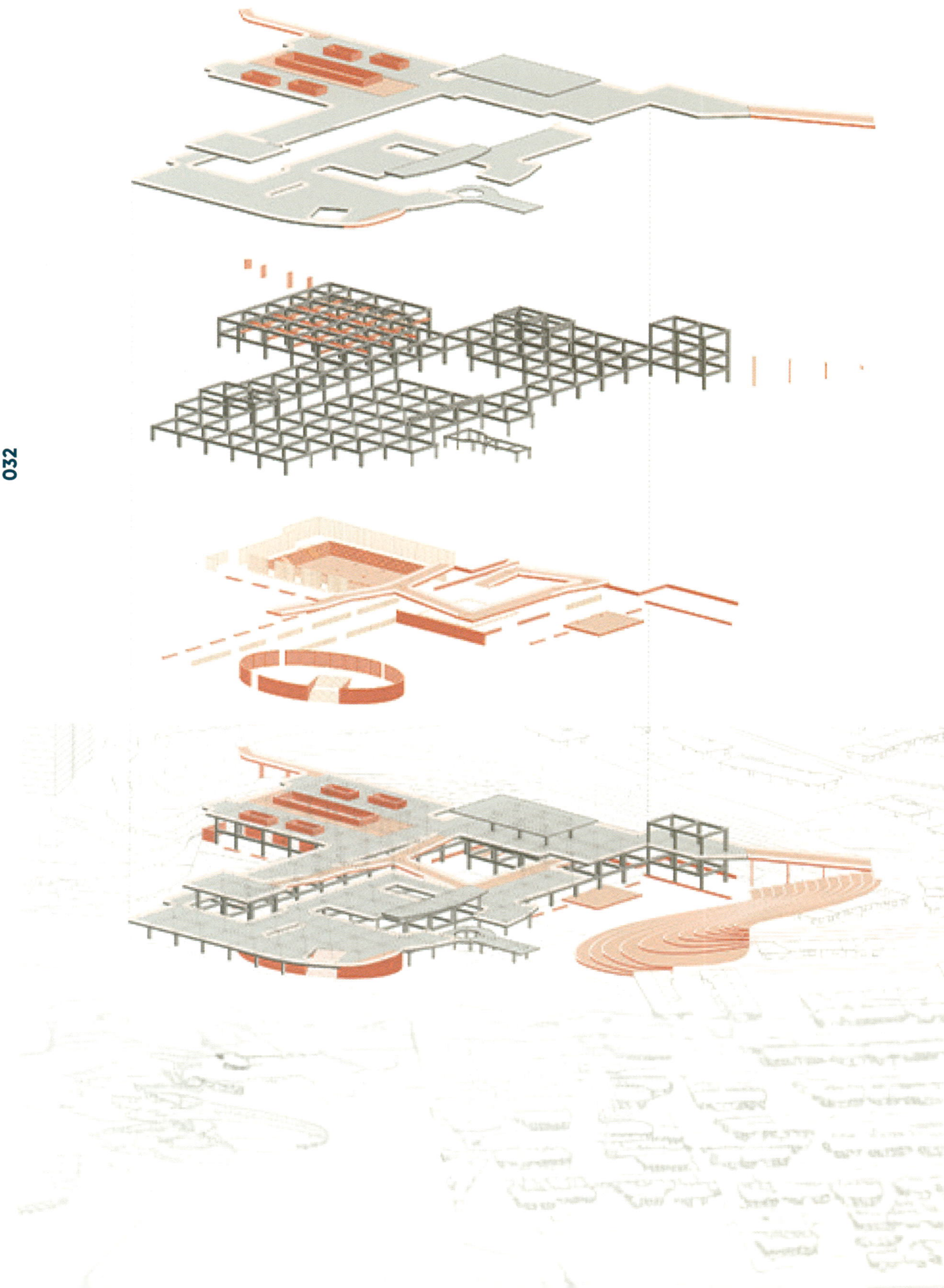

Yongsan city

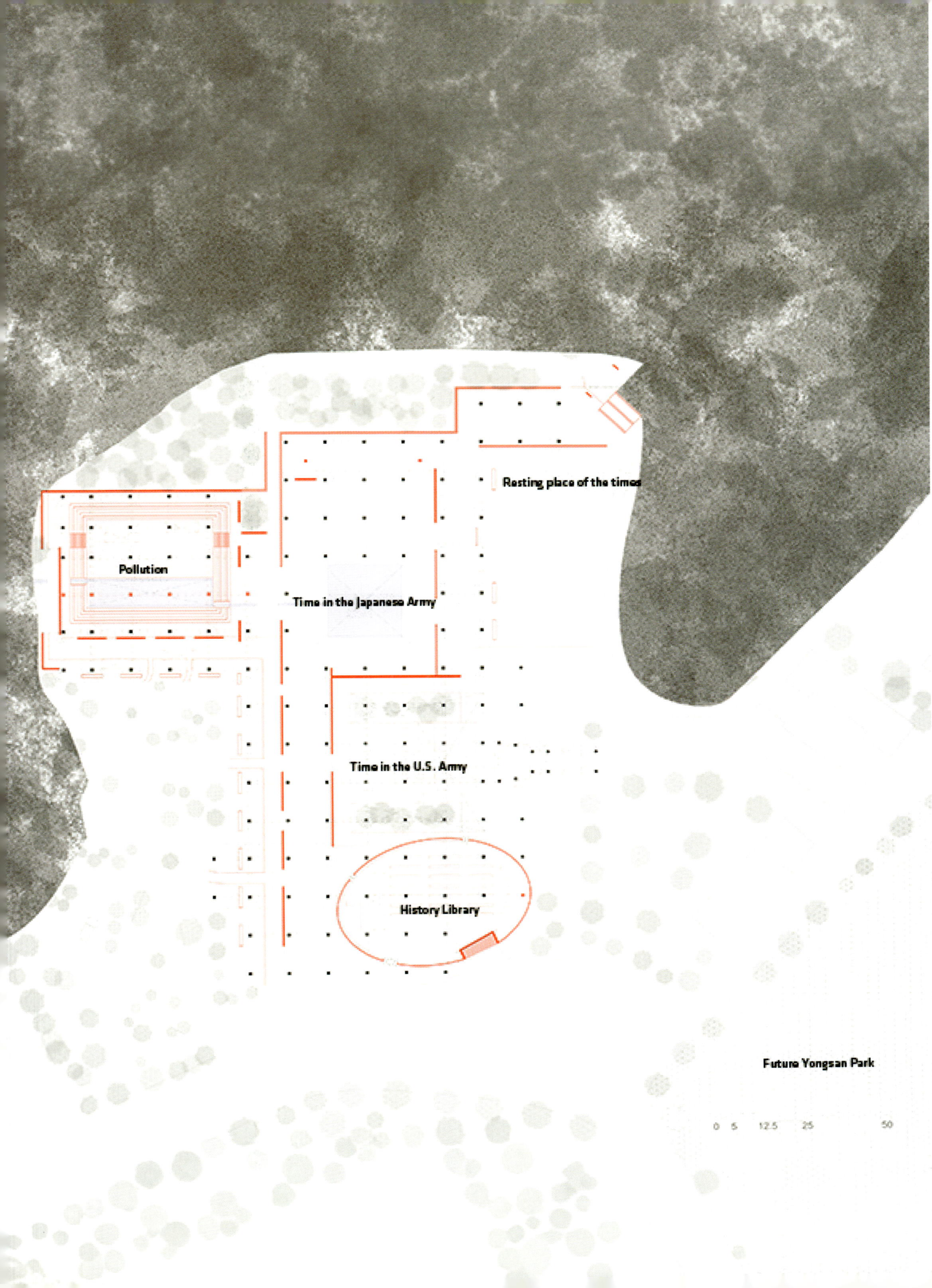
Resting place of the times
Pollution
Time in the Japanese Army
Time in the U.S. Army
History Library
Future Yongsan Park
0 5 12.5 25 50

“선도했던 땅, 동력을 잃은 땅, 복기할 땅”

열리게 되는 용산기지

일본군과 주한미군이 사용하던 군사 기지로 쓰였던 이 땅은 최근 미군이 용산기지 반환을 진행하고 있다.
이로 인해 기지 일부가 점차 개방되고 있다. 용산기지 내부와 외부는 명확한 경계로 구분되어 왔기 때문에 두 지역 간의 시간적 분위기는 확연하게 차이가 나고 있다. 2023년의 현대적인 용산 도시와 달리 용산 기지 내에는 1960년대 분위기를 유지하고 있다.

현재 South Post 서쪽 일대의 부지 중 일부는 이미 반환되어 어린이 정원으로 활용되고 있다. 그러나 아직 개방되지 않은 부지도 존재한다.
반환부지는 용산도시, 어린이정원, 미군병영, 일본군의 흔적, 미래 용산공원이 서로 겹쳐져 각각의 고유한 시간적 흐름이 교차하는 지점이다.
이 반환 부지 중에서도 미군 병원은 여러 시간축이 가장 잘 겹쳐질 수 있는 중심점에 존재하고,
병원 건물은 South Post의 상징적인 건물 중 하나이기에 이 병원부지를 사이트로 선정하게 되었다.

병원 부지는 용산기지에 속해 있어 닫혀 있던 기간 동안 1960년대의 분위기를 유지하고 있지만, 실제로 병원 자체는 여러 변화가 있어 왔다.
특히, South Post에서 가장 큰 변화를 겪은 부지 중 하나가 바로 병원 부지이다. 현재 반환 부지 중 어린이 정원을 포함한 부분은 시간에 따른 큰 변화가 없어 고유의 시간축을 보존하고 있다. 그러나 병원부지는 기지 내부와는 달리 지속적으로 변화해왔다.
병원부지는 중첩된 다양한 시간축을 가지고 있기 때문에, 이 곳이 기지의 내외부 시간축 간의 경계를 연결하는
역할을 할 수 있다고 판단했다.

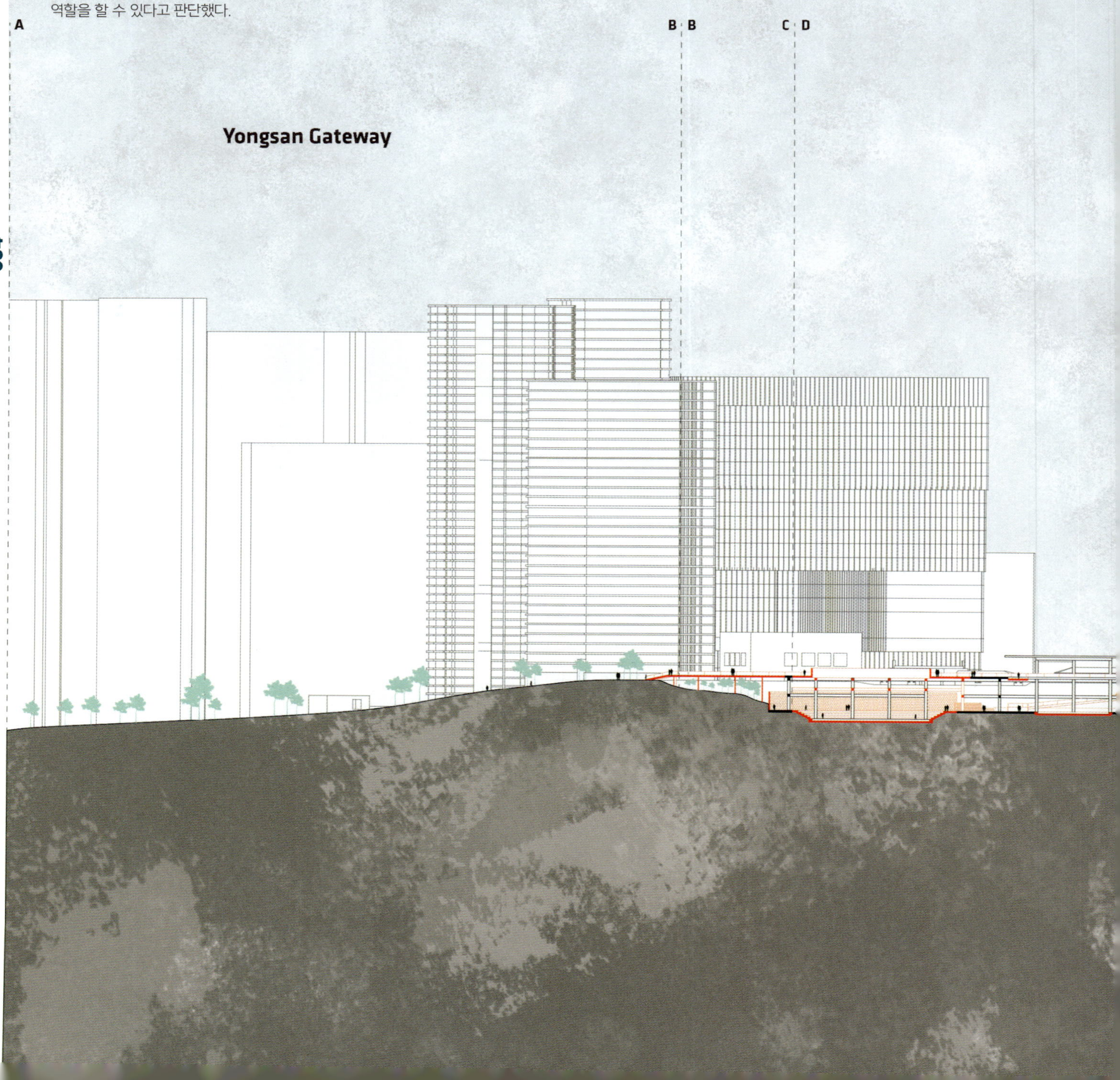

선도하는 공간에서 되돌아보는 공간으로

병원 부지는 역사적으로 시대를 선도해왔다. 일제 시기에는 용산 주변의 건물보다 자본과 기술력 면에서 더 높은 수준의 조선총독청사가 있었으며, 미군이 점유할 때에는 선진 의료체계와 건설 기술을 도입하여 건물을 짓고, 지속적인 증축과 리모델링을 진행해왔다.
그러나 미군이 부지를 반환하면서 병원을 평택으로 이전하였고, 이 공간은 닫혀 있고 비어진 공간이 되었다. 이전에는 선도적인 역할을 해왔지만, 현재는 그 동력을 잃었다.

이 동력을 다시 되찾아주는 것은 사실상 불가능하기에 병원 부지에서는 일어났던 역사를 다시 되집어보는 공간으로 조성하고자 했다.
역사성을 갖추기 위해 병원을 완전히 해체하여 새로운 것을 짓는 것 보다는 과거 미군 병원의 흔적을 남겨두는 것이 중요하다고 생각했다.
따라서 이 공간에는 과거 미군 병원이 자리하던 흔적을 존중하고 강조하여 공간 자체가 과거의 역사와 연결된 증거를 지니도록 하여, 과거로부터의 시간적 흐름을 가져오려 했다.

병원은 폐쇄적인 프로그램을 가진 공간이지만, 이제는 반환되어 프로그램이 변해야 되는 시점이다.
심지어 용산기지는 "용산공원 조성 특별법"까지 제정되어 반환된 땅은 공원으로 조성될 예정이다.
따라서 이 공간은 공원화 및 공공화할 필요가 있다.

그래서 이 공간을 미래에는 열린 공간으로 만들어야 한다. 단절되어 있던 베를린 장벽을 허물어 사람들이 교류하는 것처럼 폐쇄적인 메스에 있는 벽을 허물어 내고자 한다. 지금은 외부로부터 닫혀 있지만, 미래에는 벽을 허물어내어 건물 자체가 하나의 역사를 느낄 수 있는 역사 공원으로 변화하여 시간적 차이를 연결하는 공간을 만들고 싶다.

**Yongsan Stataion
& Chilren's Garden**

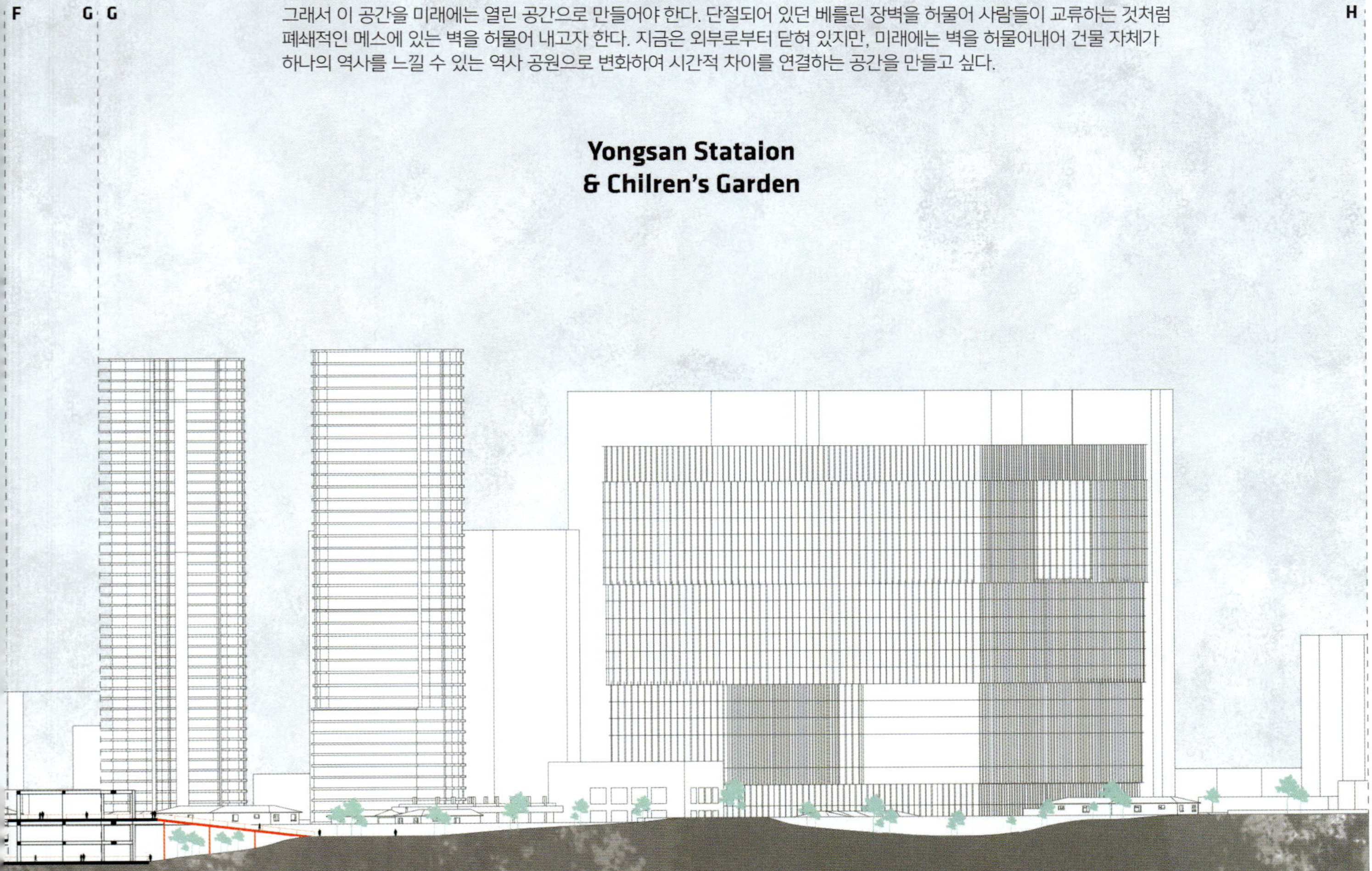

Yongsan Stataion

역사를 기록하고 답습하는 역사공원

역사를 간직하고, 사람들에게 역사적 차이를 연결 해주는 곳인 만큼 이 공간은 이전의 역사를 기록하고 답습할 수 있는 느슨한 형태의 열린 도서관으로 조성하고자 한다. 열린 큰 공간을 담장을 통해 구획하고 각각의 역사공간으로 구성하여 사람들이 편하게 역사를 접하게 만드는 것이다.

그리고 공간을 나눌 때는 부지 내의 있었던 역사를 구분하여 나누고자 했다.
오염, 일본군, 미군, 자연에 대해 다루고, 각 공간에 맞게 사람들이 향유하고, 독서하면서 공원을 즐기는 모습을 상상했다.

공간을 나누는 벽은 현재 어린이정원에 있는 투과성 있는 담장과 유사한 형태를 사용하여 공간 자체를 구획하지만 단절되지 않게 구성하고자 한다. 그리고 이러한 담장 규격을 조정해 담장 자체가 하나의 야외 사고가 될 수 있도록 조정하여 사람들이 자연스럽게 책을 읽고 사색하는 공간이 만들고 싶다.

Ichon Stataion

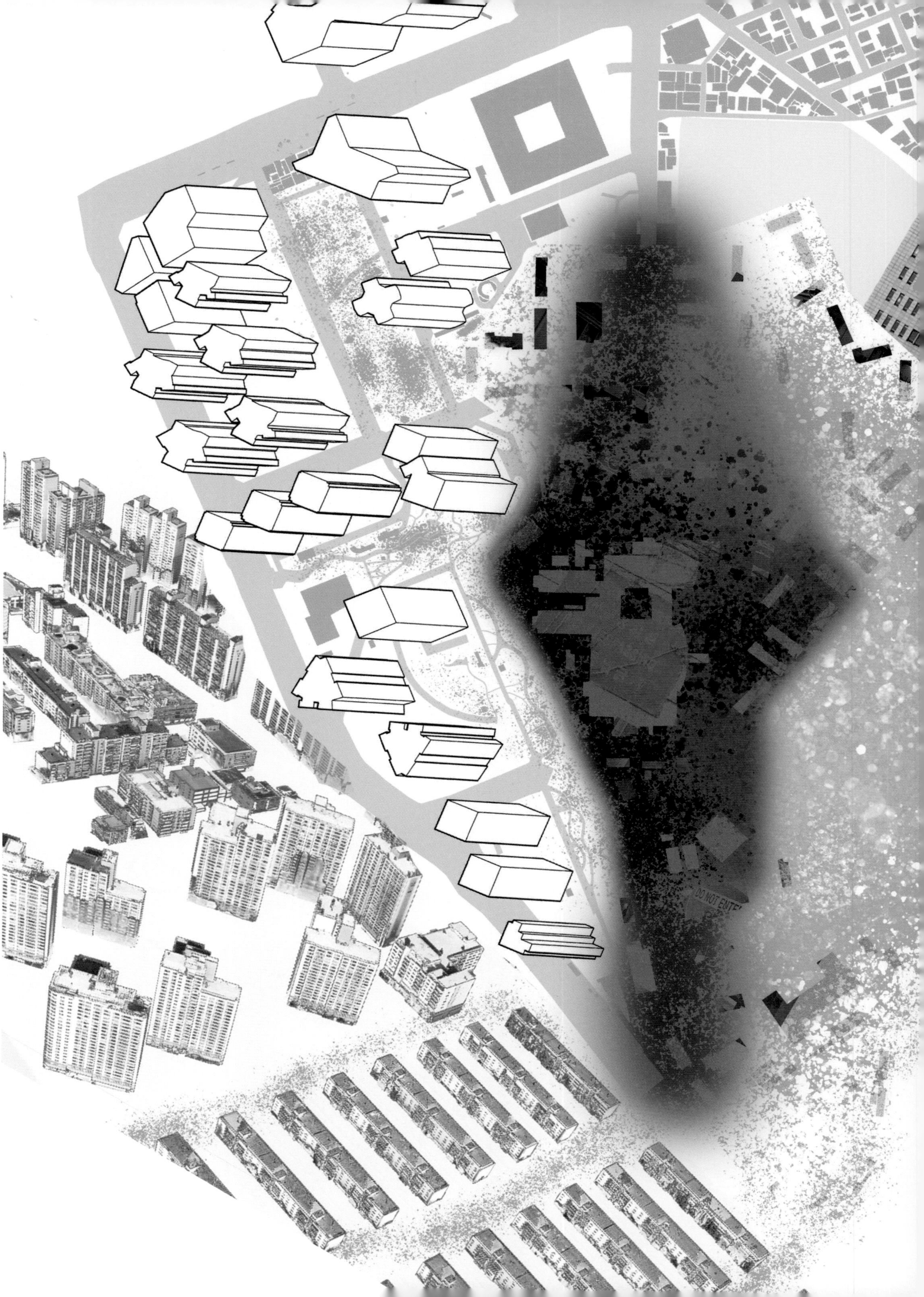

Boundary Layer(경계층)

경계와 도시

도시와 공원은 공존하며 필연적으로 그 사이에 경계가 있다. 용산기지 내부에 위치한 대상지는 현재 사용되지 않는 미군 병원 부지이나, 용산 공원이 조성됨에 따라 인공과 자연이 만나는 경계가 될 예정이다. 프로젝트를 통해 정의하는 경계층은 경계선과 달리 하나의 영역을 만든다. 경계층을 경험하는 대상은 공원 이용자 그리고 대지를 투과하는 물이다. 물은 식생으로 흡수되어 느린 정화의 과정을 거치며 사람은 경계층에 다가가며 시각, 동선상으로 이벤트를 겪는다. 물과 사람 모두 경계층에 수직방향으로 다가갈때 그 속도가 느려지며 수평방향으로 이동할때 속도가 빨라진다. 일반적인 가장자리로써 기능하는 경계가 아닌 두께를 가진 경계층을 디자인하는 것이 프로젝트의 목표이다.

시설과 기계

물과 사람이 경계층을 경험할 수 있도록 만드는 것은 사이트에 설치된 시설과 기계이다. 시설에는 우수를 모아서 바닥의 페이빙을 통해 흐를 수 있도록 하는 것이 있고, 지하수가 흐르도록 하는 파이프도 존재한다. 바닥의 기다란 녹지 라인을 형성하는 것 또한 시설과 허브가 기반이 된다. 시설은 축을 형성하는 동시에 동선을 보조하고, 사람들이 휴식할 수 있는 공간을 제공하며 물을 기계적으로 정화할 수 있게 한다. 시설에 수반된 기계장치로는 지하수를 끌어올릴 수 있는 펌프 그리고 모아진 물을 물리적, 화학적으로 정화할 수 있는 필터가 있다.

식물과 정화

물은 기계 장치를 통하는 것 뿐만 아니라 식물을 통해서도 정화된다. 여기서 정화는 용산 기지 내부 토양과 지하수의 오염에 대한 것이다. 기지 내부의 오염된 물과 토양은 정화과정을 거쳐 기지 외부로 나가기도 하며 기지에서 그대로 사용되기도 한다. 경계층이 하나의 영역을 만들어 인공과 자연의 만남을 의도하더라도 사람들이 사용하기에 오염도가 너무 높다면 적극적으로 사용될 수 없을 것이다.

정화 중에서도 식물을 통한 정화(phytoreme diation)는 중금속을 흡수할 수 있는 식물종을 활용하게 된다. 몇몇의 연구 결과에 나와 있듯이 버드나무, 해바라기, 말냉이, 겨자식물, 수크령이 중금속 흡수 및 저장을 할 능력을 갖추고 있다. 이러한 식물 정화와 시설등을 통해 실제 정화 되는 것 뿐 아니라 사람들로 하여금 용산 기지가 기름 유출로 인한 오염으로 뒤덮인 토양으로 남아있는 것이 아니라 회복되어 가고 있는 공원이라는 인상을 남길 수 있다. 이 경계층은 현재의 기지가 가지고 있는 오염에 대한 이야기 뿐 아니라 앞으로 다가올 오염에도 도움이 된다.

기지 내부는 토양 오염에 따라서 지하수도 오염되어있는 상황이다. 경계층에서 끌어올려진 오염된 지하수는 정화의 기능을 하는 식물에 뿌려져서 중금속이 제거될 수 있도록 한다. 우수와 지하수가 토양에 뿌려지고 식물에 흡수되며, 일부는 시설로 이동해 정화되는 하나의 시스템을 구축하는 것이 목표이다.

Boundary and the city

The city and the park coexist, and there is inevitably a boundary between them. The site, located inside Yongsan Garrison, is the unused US military hospital, but with the creation of Yongsan Park, it is expected to become a boundary where artificiality and nature meet. Unlike the boundary, the Boundary Layer defined through the project creates an area. Those who experience the Boundary Layer are park users and water permeating the land. Water is absorbed into vegetation and goes through a slow remediation process, and people experience events visually and movingly as they approach the Boundary Layer. Both water and people slow down as they approach the Boundary Layer vertically and increase in speed as they move horizontally. The goal of the project is to design a Boundary Layer with a thickness rather than a boundary that functions as a normal edge.

Facilities and Machinery

It is the facilities and machines installed on the site that make it possible for water and people to experience the Boundary Layer. The facilities include those that collect rainwater and allow it to flow through the bottom paving, and there are also pipes that allow underground water to flow. The facilities and hubs form the long greenery at the bottom as well. The facilities form axes, while also assisting in the flow of traffic, providing a resting place for people, and allowing the mechanical purification of water. The mechanical devices include pumps that can raise underground water and filters that can purify the collected water physically and chemically.

Plants and Remediation

Water is purified not only through machinery but also through plants. Remediation here is about the contamination of soil and underground water inside the Yongsan garrison. Contaminated water and soil inside the garrison go through the purification process and are sometimes used as they are inside the garrison. Even if the Boundary Layer is intended to create an area to meet artificiality and nature, it cannot be actively used if the pollution level is too high for people to use.

Remediation through plants, among other things, utilizes plant species that can absorb heavy metals. As some studies have shown, willow trees, sunflowers, horseradish, mustard plants, and sucrose have the ability to absorb and store heavy metals. In addition to being actually purified through such phytoremediation and facilities, it can leave the impression that Yongsan Garrison is a recovering park rather than remaining as contaminated soil from oil spills. This Boundary Layer is helpful not only to talk about the pollution that the garrison has, but also to come.

Inside the garrison, underground water is also contaminated due to soil pollution. Contaminated underground water raised from the Boundary Layer is sprayed onto the plants that function as purification so that heavy metals can be removed. The goal is to build a system in which rainwater and underground water are sprayed into the soil and absorbed by plants, and some of them are moved to facilities to purify them.

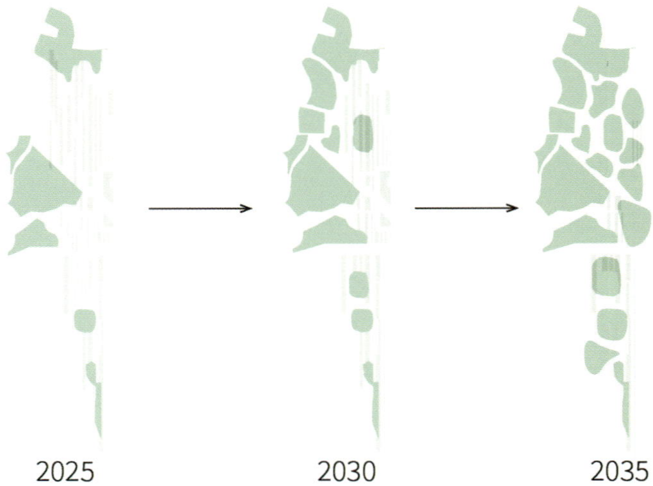

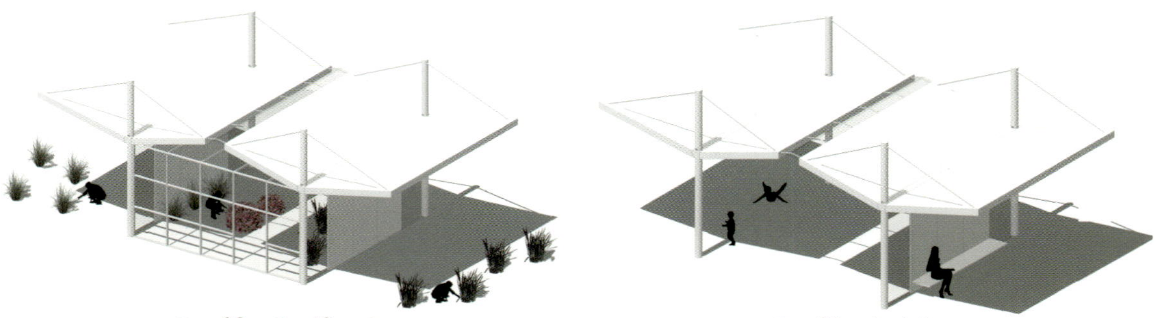

Roof for Purification

Roof for Public

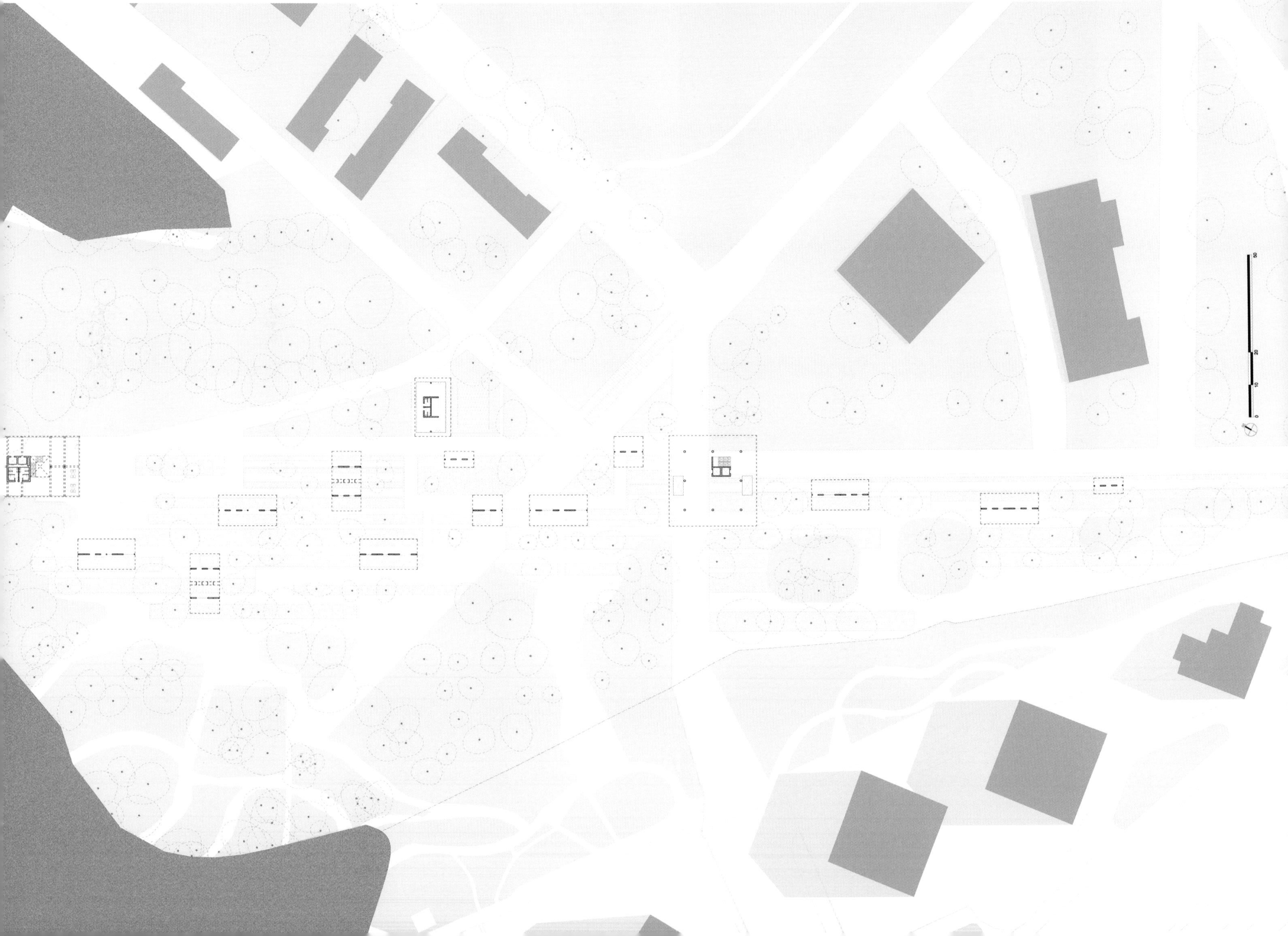
0
10
20
50

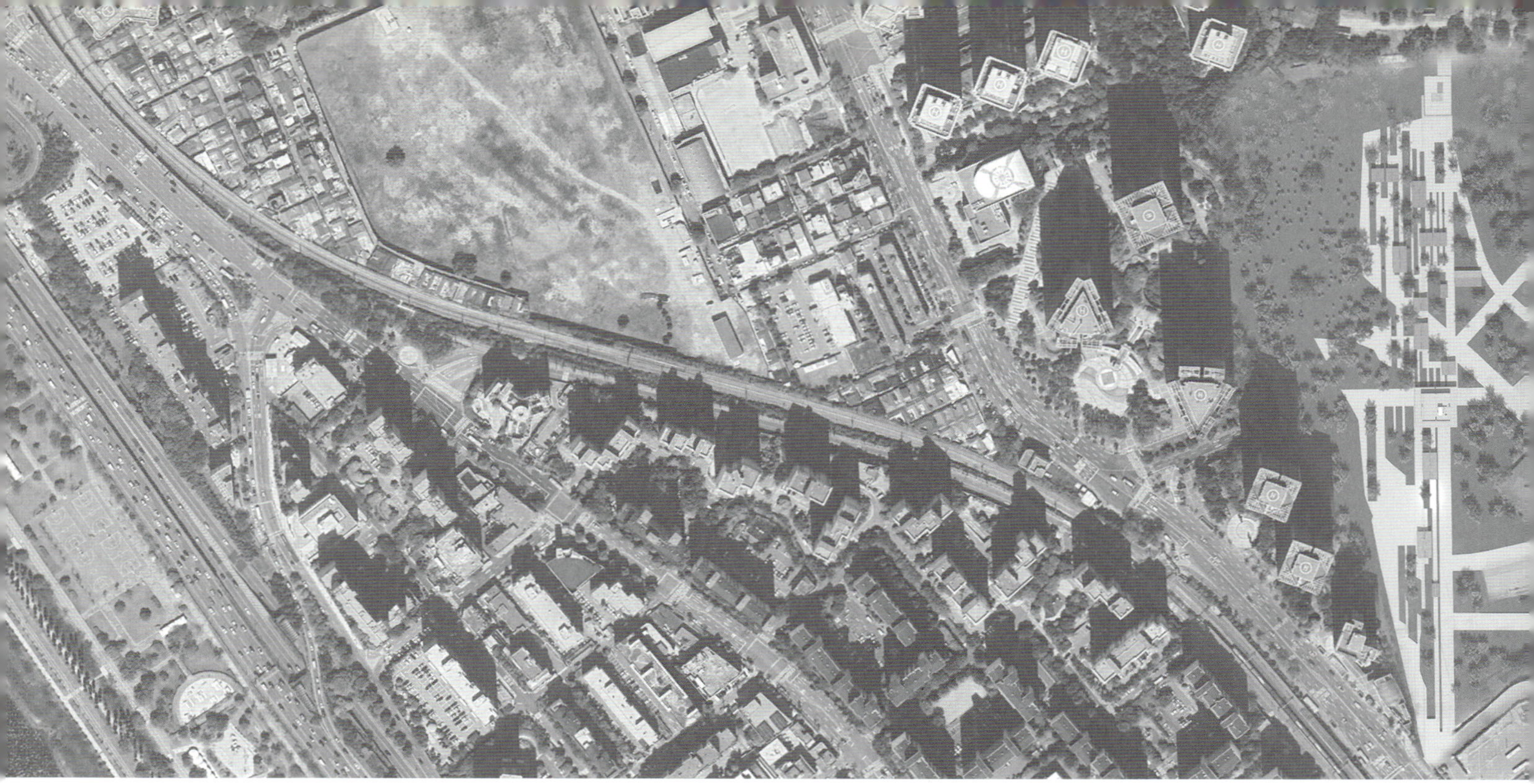

1:200, Acrylic & Cork

YONGSAN

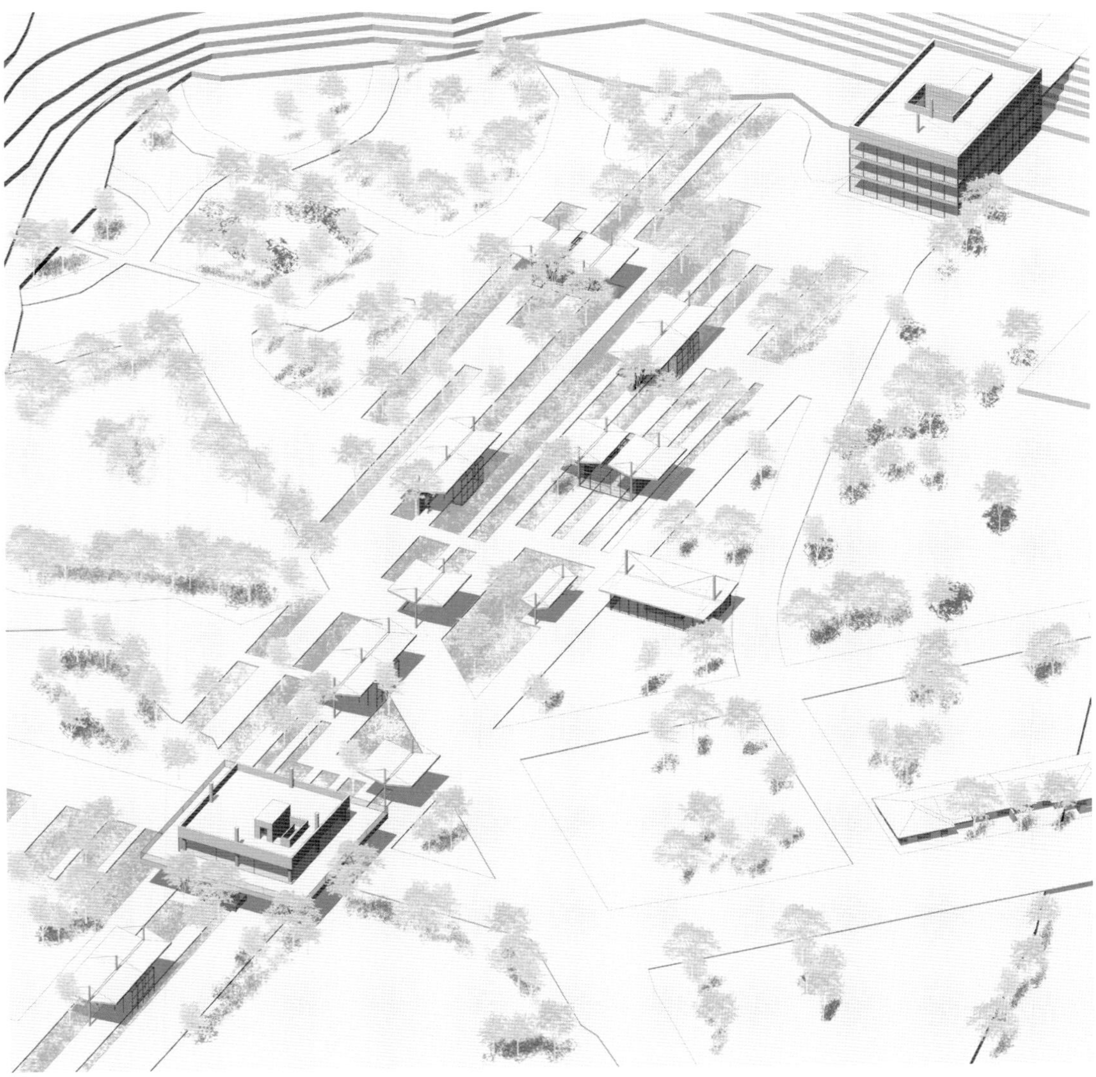

THE ARCHITECTURE OF AMERICAN HEGEMONY IN SEOUL

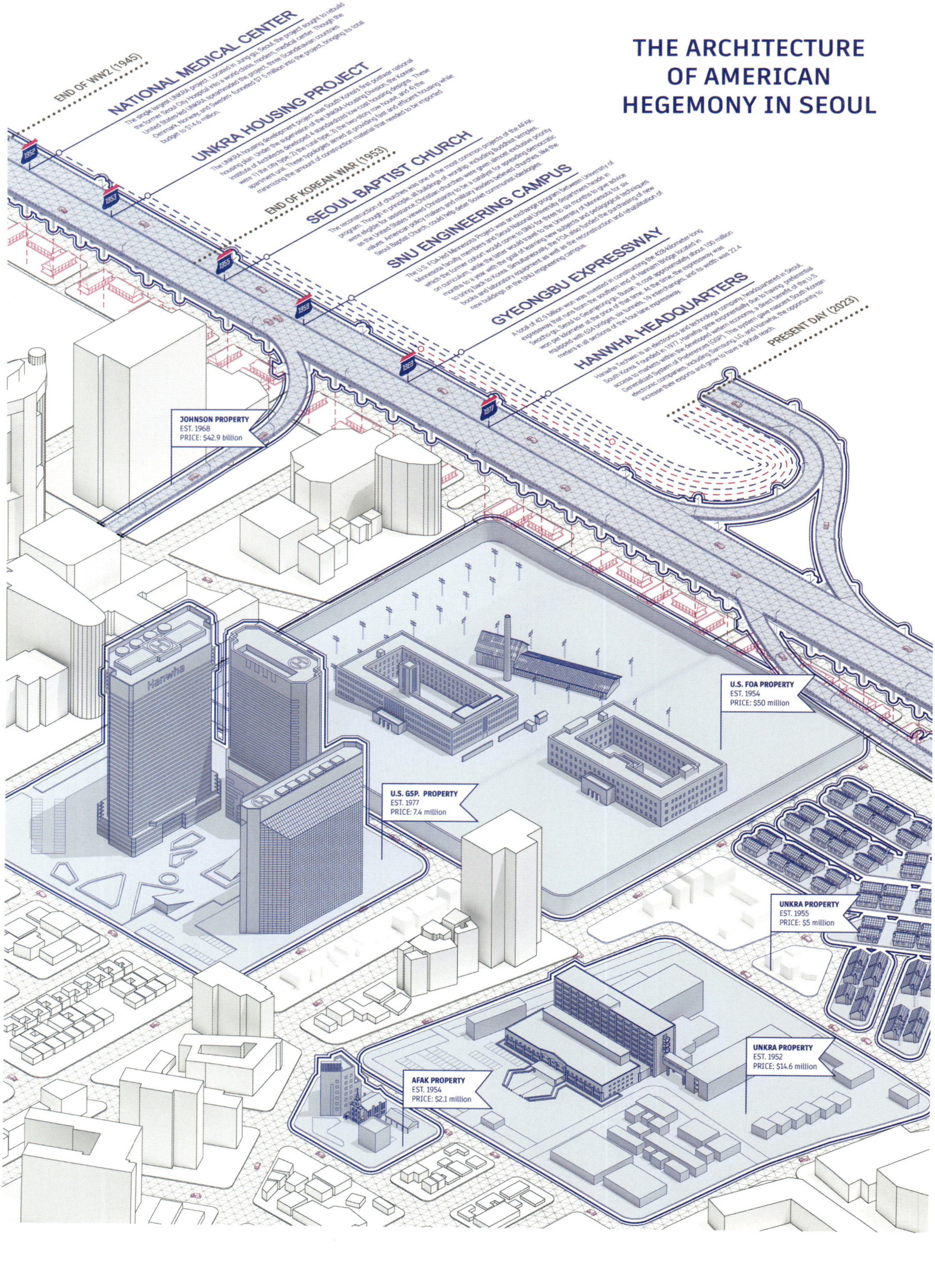

046

WATER POLLUTION

Date of Incident: 2000 - 2005
Location: Morgue with in Yongsan Garrison, Yongsan-gu, Seoul
Actors: United States Forces Korea, Green Korea United

Description:
In February 2000, an American civilian employee at Yongsan Garrison disposed of 470 bottles of formaldehyde down a sink in the morgue. This was done under the orders of the deputy chief of the morgue, Albert Mcfarland. The incident involved the improper discharge of toxic chemicals into the Han River system and was made public in July by the civilian employee. Lt. Gen. Daniel Petrosky of the 8th U.S. Army Headquarters officially apologized for the illegal discharge, admitted to violating Korean domestic law and U.S. military regulations, and announced a review of environmental programs for U.S. Forces in Korea on September 8. Mcfarland received a thirty-day pay reduction as discipline.

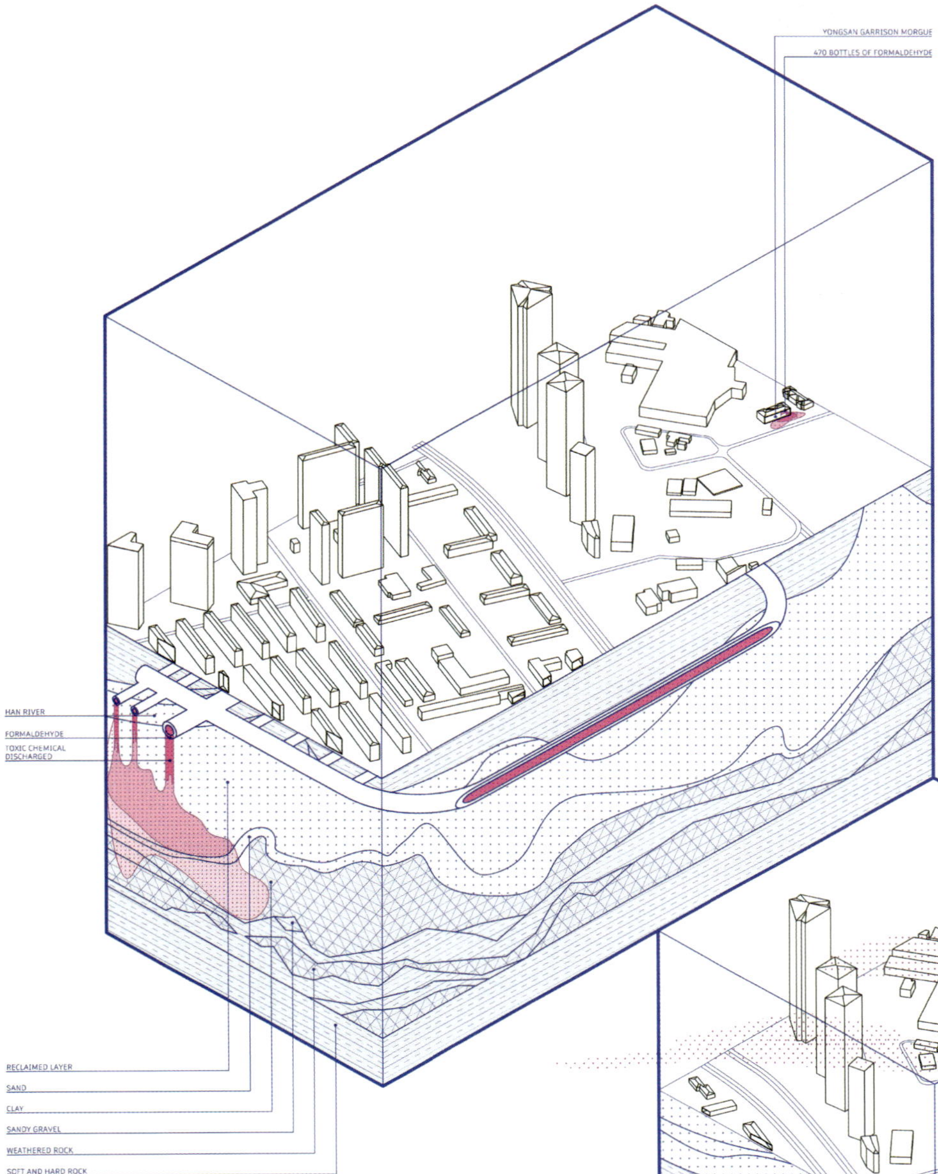

AIR POLLUTION

In East Asia, particularly in Seoul, South Korea, transboundary air pollution worsens during the fall and winter due to prevailing west winds, known as "the westerlies." During these seasons, South Korea experiences persistent winds carrying pollution from China, leading to increased air pollution levels. A study conducted from 2013 to 2017 in Seoul found that a mere 1 microgram per cubic meter increase in PM2.5 (fine particulate matter) resulted in elevated mortality rates. Specifically, this increase equated to an additional 31.2 deaths per million people annually. Infants were the most vulnerable, experiencing a 2.1% increase in mortality, while individuals with respiratory and cardiovascular diseases saw a 1.1% increase in mortality.

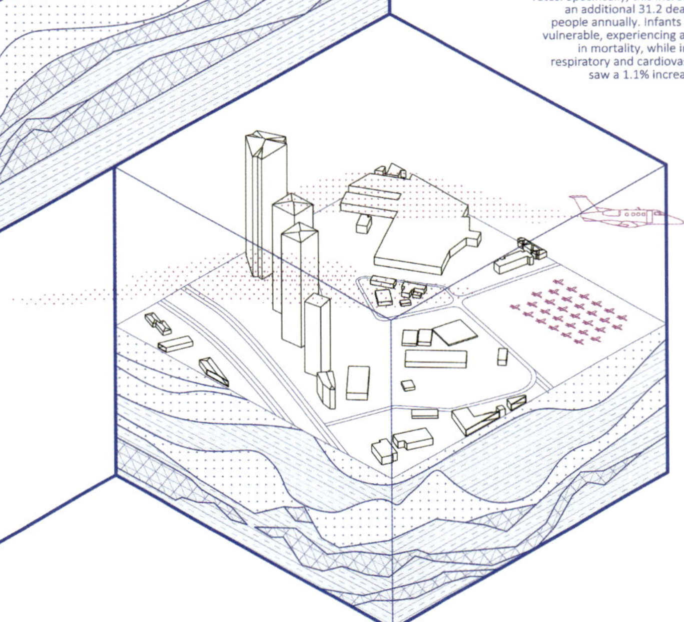

SOIL CONTAMINATION

Site: Noksapyeong subway station, Yongsan-gu, Seoul
Date of Incident: 1988- Ongoing
Location: Noksapyeong subway station, Yongsan-gu, Seoul
Actors: Korea Rural community & Agriculture Corporation, Seoul Metropolitan Government, Ministry of Environment, United States Forces Korea

Description:
In January 2001, groundwater at Noksapyeong station was contaminated by a significant oil spill, emitting a strong flammable odor. The spilled oil saturated absorption cloths placed in the water. In February and March, the Seoul Metropolitan Government collected samples from subway tunnel manholes and determined that the oil came from a gas station in Yongsan Garrison.

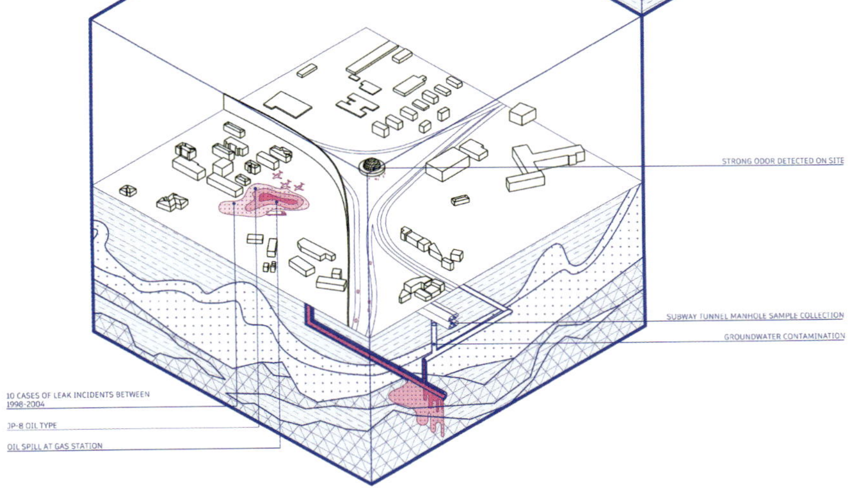

ARCHIPELAGO NETWORKS OF SEOUL:

Ecology, Naitive Species, and Interconnectivity

The complexity and variety of ecosystems formed by the diverse climate and complicated topography affect the biodiversity which inhabits the Peninsula. Sub-alpine coniferous forests are common in the northern region, deciduous broadleaf forests are common in the central region, and warm, temperate evergreen forests are common in the southern and island regions. The natural conditions and variety of vegetation also create variations in ecosystem productivity, resulting in distinct micro-habitats for a wide diversity of faunal communities.

Sources

SHIFTING ARCHIPELAGO:

Base Closure, Relocation, Expansion, Consolidation

"For it is at the precise moment of incipient prepositioning or demobilization—mirrored movements described the same hackneyed adage that 'in time of piece prepare for war'—when conflict is displaced and dormant rather than confused by the frictions and the fogs of war, that we learn most about the magnitude of power conveyed spatially and temporally through military facilities and installations." (Belanger & Arroyo, Ecologies of Power, 13).

This catalog does not include 17 sites indisclosed in area or location. Most installation borders were crowd-sourced in Wikimapia. Dates of installation closures are recorded according to the Land Partnership Plan and Yongsan Relocation Plan and do not reflect ongoing delays and negotiations.

Sources

David Vine. "Base Nation." Base Nation. Accessed September 25, 2023. https://www.basenation.us/.

Pierre Belanger and Alexander Arroyo. Ecologies of Power: Countermapping the Logistical Landscapes & Military Geographies of the US Department of Defense. Cambridge, MA: MIT Press, 2016.

US Department of Defense. "Base Structure Report, Fiscal Year 2022." 2022. https://www.acq.osd.mil/eie/Downloads/BSI/Base%20Structure%20Report%20FY22.xlsx.

"Wikimapia." Accessed September 25, 2023. https://wikimapia.org/#lang=en&lat=39.949400&lon=-75.145700&z=12&m.

CURRENTLY OPERATING
CLOSED / RELOCATED
UNDISCLOSED BORDER
ARMY
NAVY
AIR FORCE
MARINE CORPS
OTHER
LAND PARTNERSHIP PLAN
LPP LAND GRANT (EXPANSION)
YONGSAN RELOCATION PLAN
1km

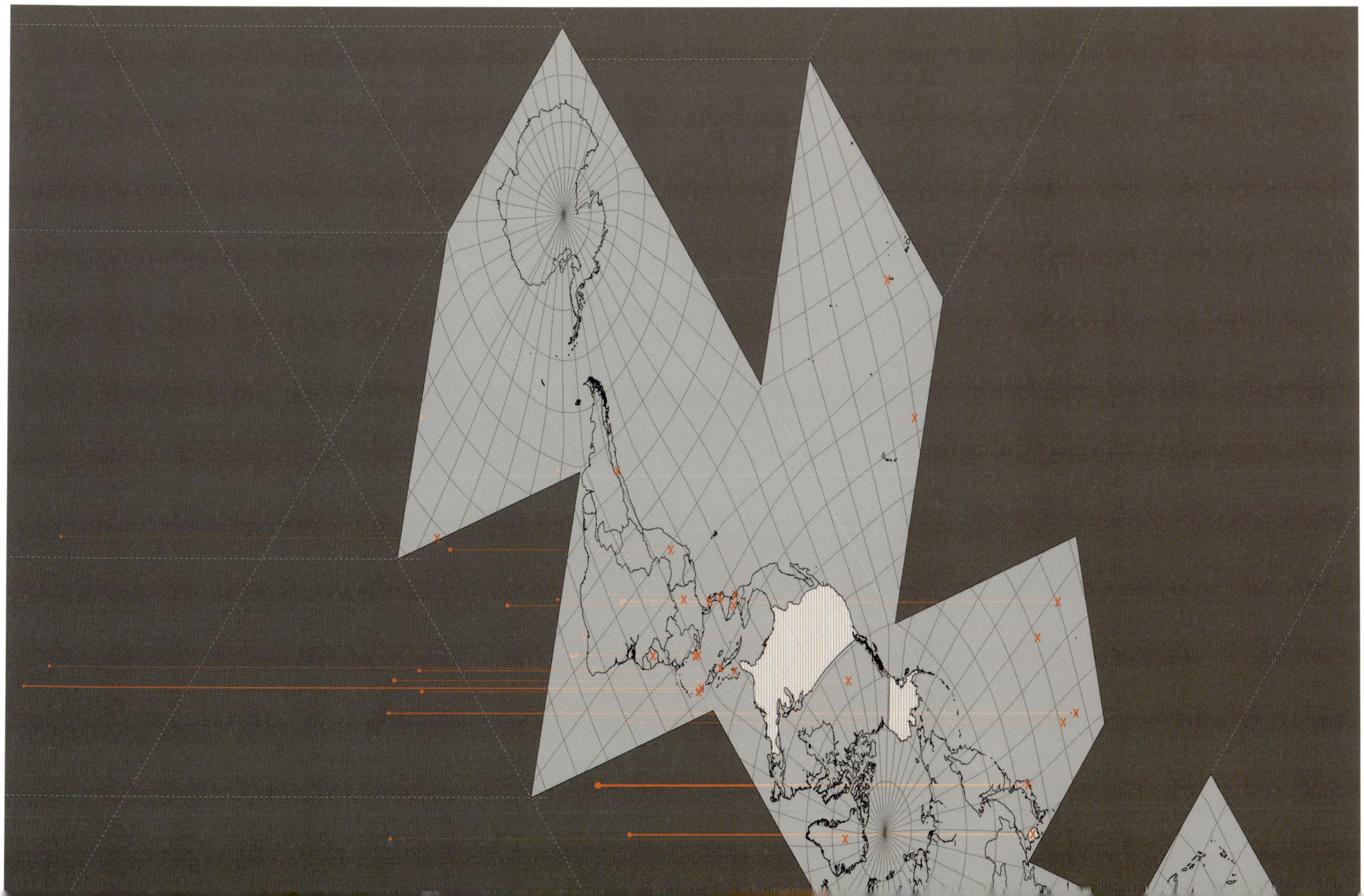

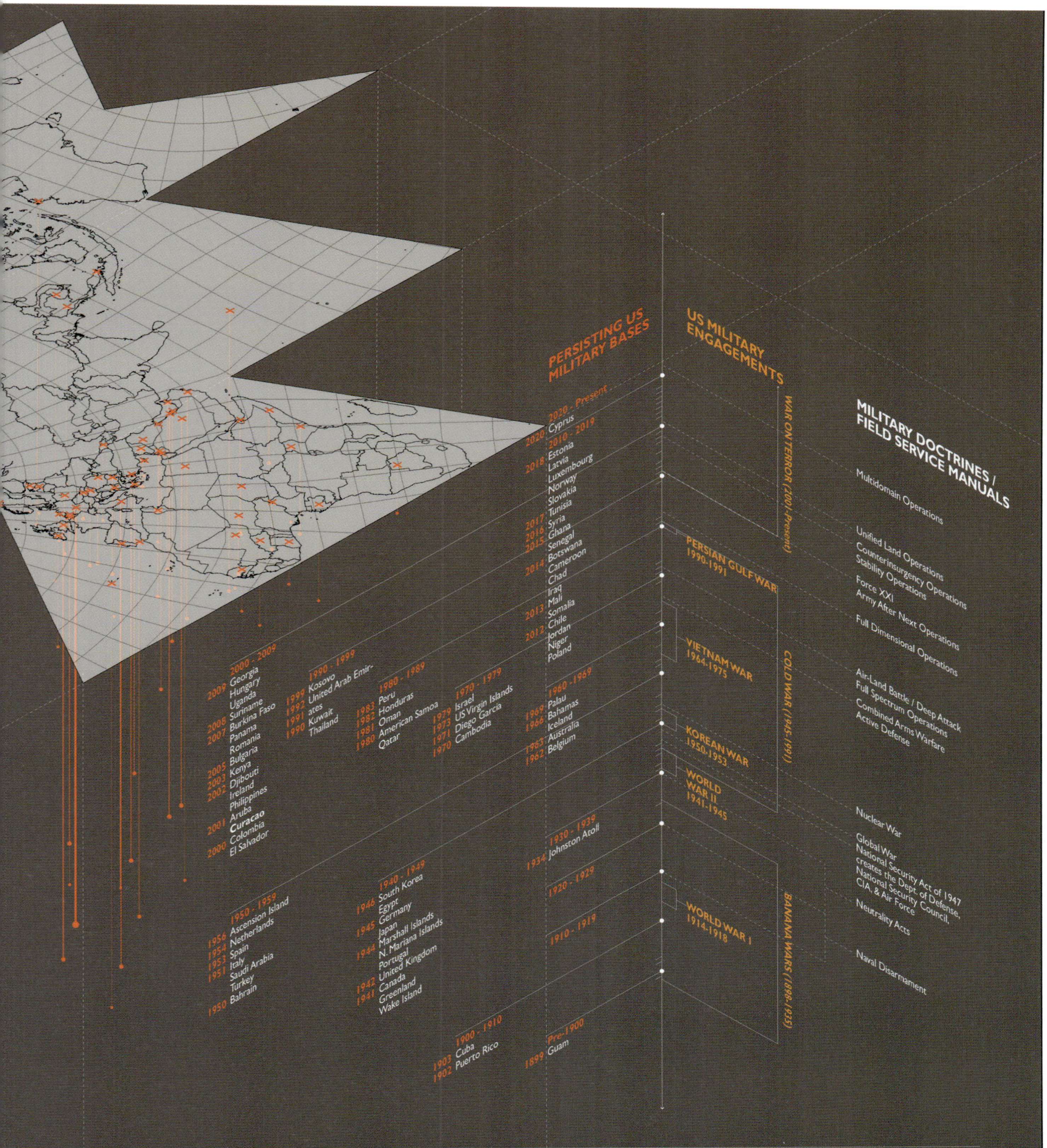

XPANDING RCHIPELAGO:

eocolonial Logics & and Militarization

"Oceanic and orbital in scale, the rise in infrastructure build-up since has been political, geographic, and logistical in kind...Considered as a globally distributed, materially diverse landscape, this infrastructural milieu both expresses, enforces and delivers the remarkably uneven forms and relations of power through which strategies of force protection and preemptive action are undertaken through an increasingly complex matrix of logistical operations and supply chains, located between objectives of force protection and preemptive action."
(Belanger & Arroyo, Ecologies of Power, 15)

According to its annual Base Structure Report, the US Department of Defense manages a portfolio of real property that includes more than 643,973 facilities and structures on 25.7 million acres of land across 4,860 sites (in the US and more than 80 US territories and foreign countries). These sites include military bases and installations, but due to sensitive and covert operations, this list is not comprehensive. Some may argue that US military presence is necessary for global security, conflict preemption, and humanitarian aid. But the sheer number, spread, and ongoing proliferation of military installations should make us consider the spatial, political, ecological, financial, and neocolonial repercussions of a military archipelago that spans the planet.

Sources

David Vine. "Base Nation." Base Nation. Accessed September 25, 2023. https://www.basenation.us/.

Jensen, Benjamin. Forging the Sword: Doctrinal Change in the U.S. Army. Stanford University Press, 2016.

Pierre Belanger and Alexander Arroyo. Ecologies of Power: Countermapping the Logistical Landscapes & Military Geographies of the US Department of Defense. Cambridge, MA: MIT Press, 2016.

US Department of Defense. "Base Structure Report, Fiscal Year 2022." 2022.
https://www.acq.osd.mil/eie/Downloads/BSI/Base%20Structure%20Report%20FY22.xlsx.

This project advocates for the ecological and cultural remediation of the formerly foreign-occupied US Army Yongsan Garrison, a site now being returned to the Korean government for public use despite ongoing concerns of residual soil contamination. Simultaneously, there is a growing concern about the increasing prevalence of lonely deaths among Koreans, a phenomenon where individuals pass away without any family or loved ones to claim their bodies or mourn their passing.

"Sacred Soil" seeks to address both challenges by reprogramming the approach the Korean government uses to handle lonely deaths. Our project envisions the establishment of a dedicated soil remediation and testing facility that operates as a funerary yard. This 3-building intervention aims not only to remediate environmental issues but also redefine traditional spaces for collective memory production and collective mourning. By integrating ecological and cultural operations, our project provides a holistic approach to historical, environmental, and social challenges, fostering compassion for lonely deaths in Korea and paving the way for a more sustainable and empathetic cultural landscape.

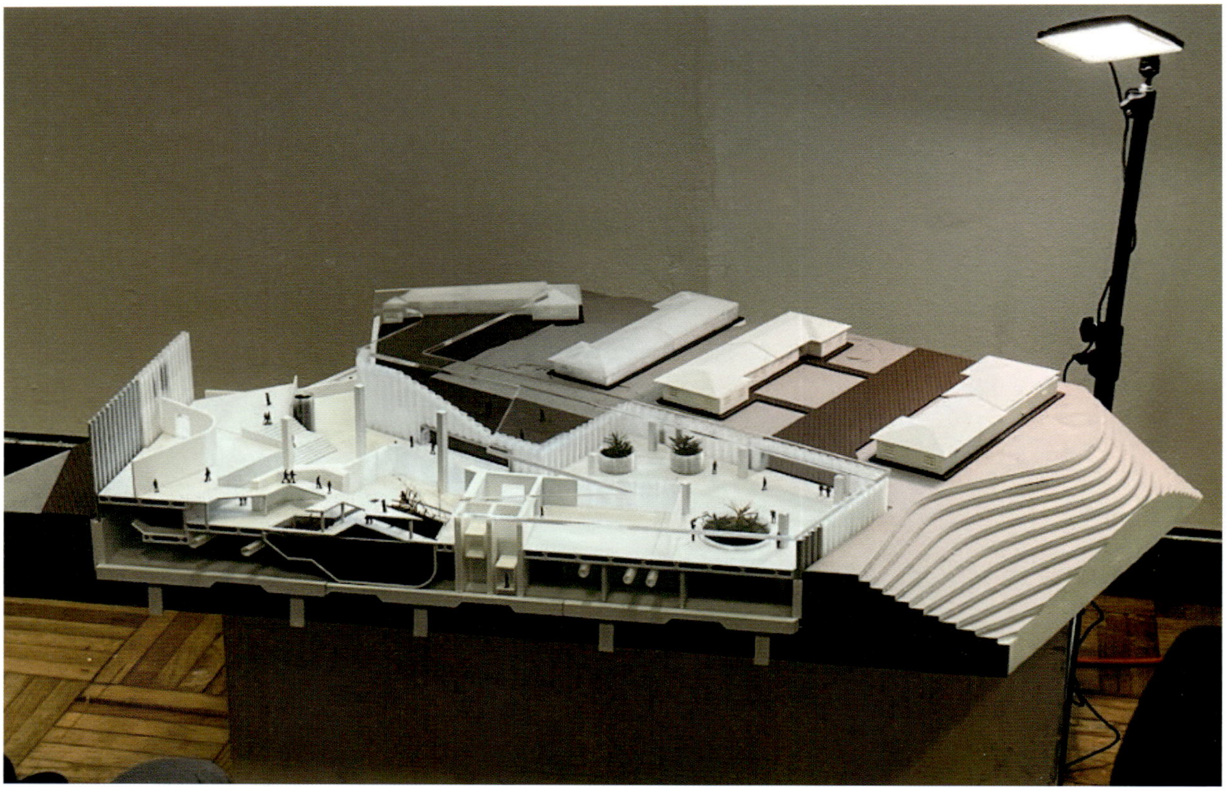

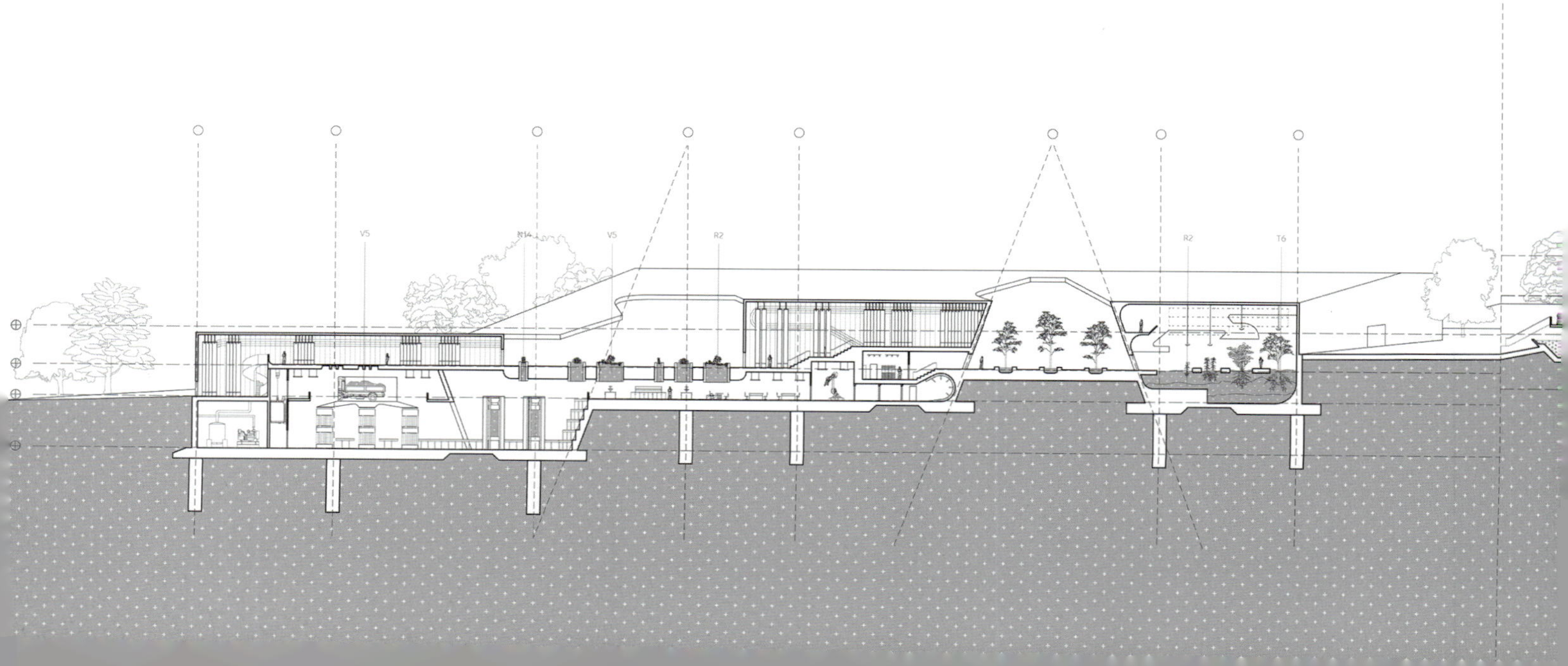

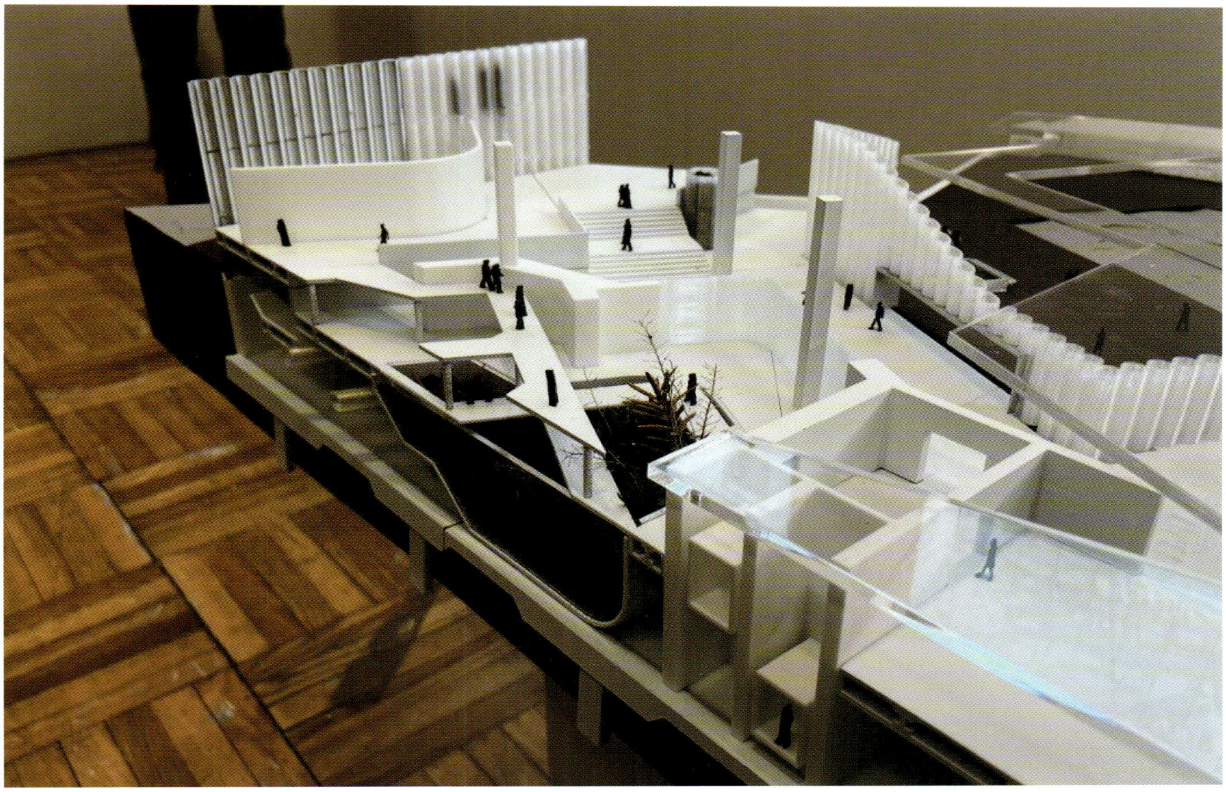

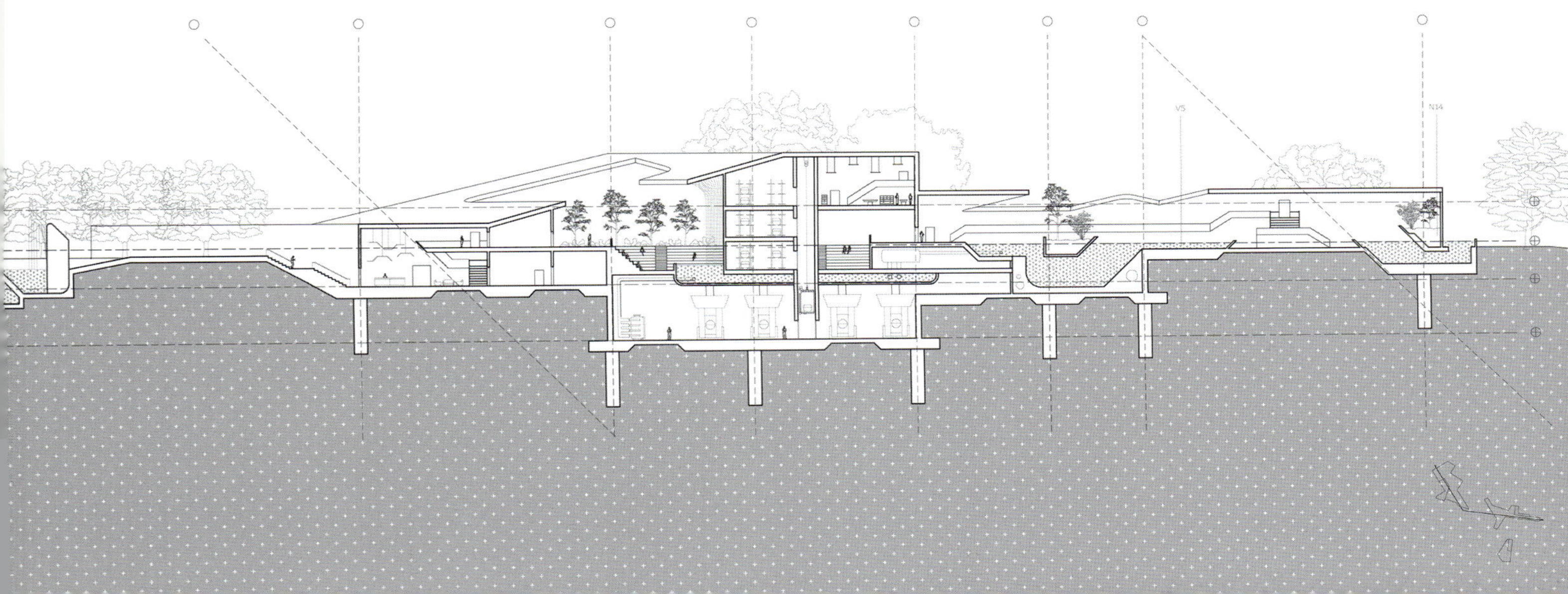

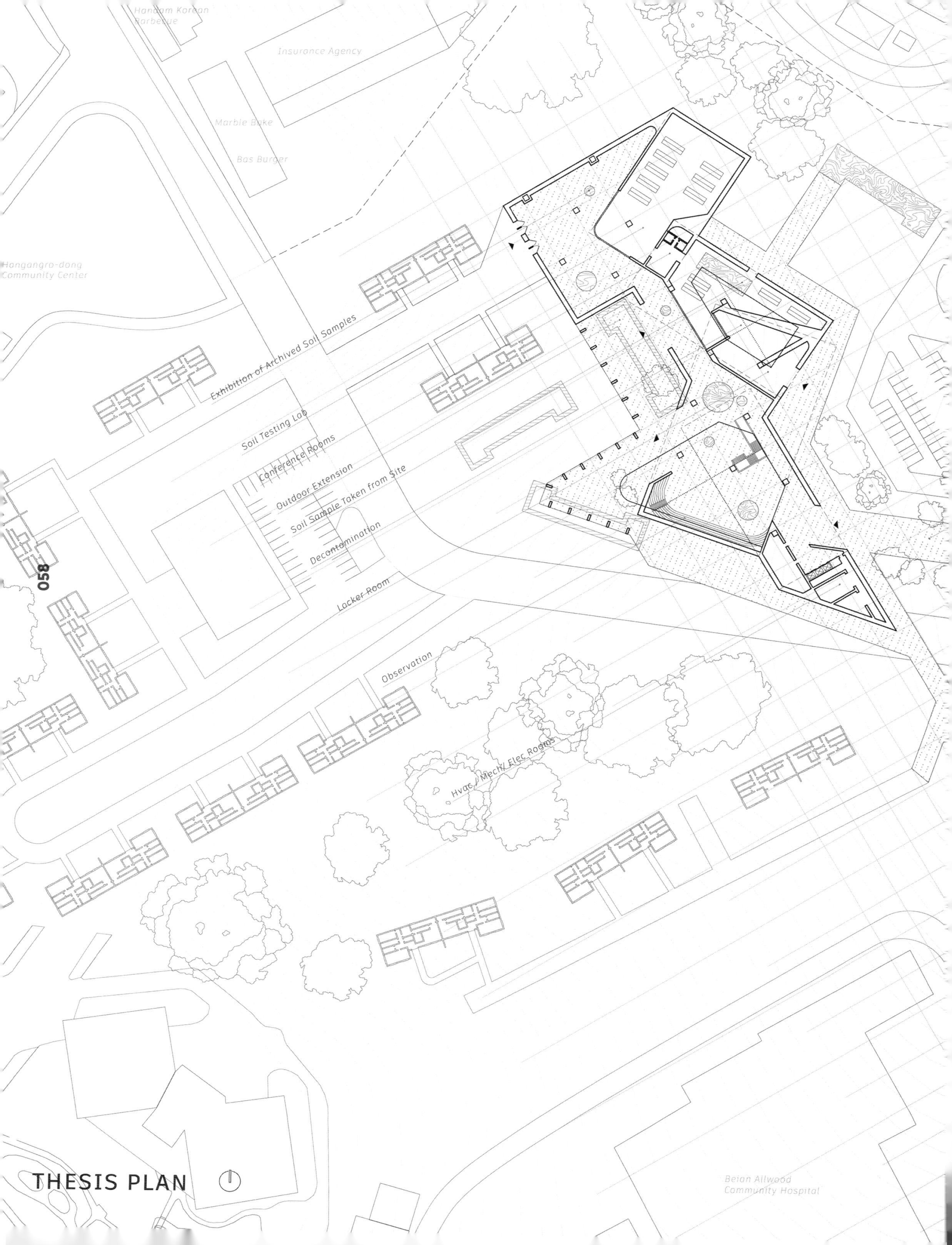
Handam Korean Barbeque
Insurance Agency
Marble Bake
Bas Burger
Hangangro-dong Community Center
Exhibition of Archived Soil Samples
Soil Testing Lab
Conference Rooms
Outdoor Extension
Soil Sample Taken from Site
Decontamination
Locker Room
Observation
Hvac / Mech/ Elec Rooms
Beian Allwood Community Hospital

THESIS PLAN

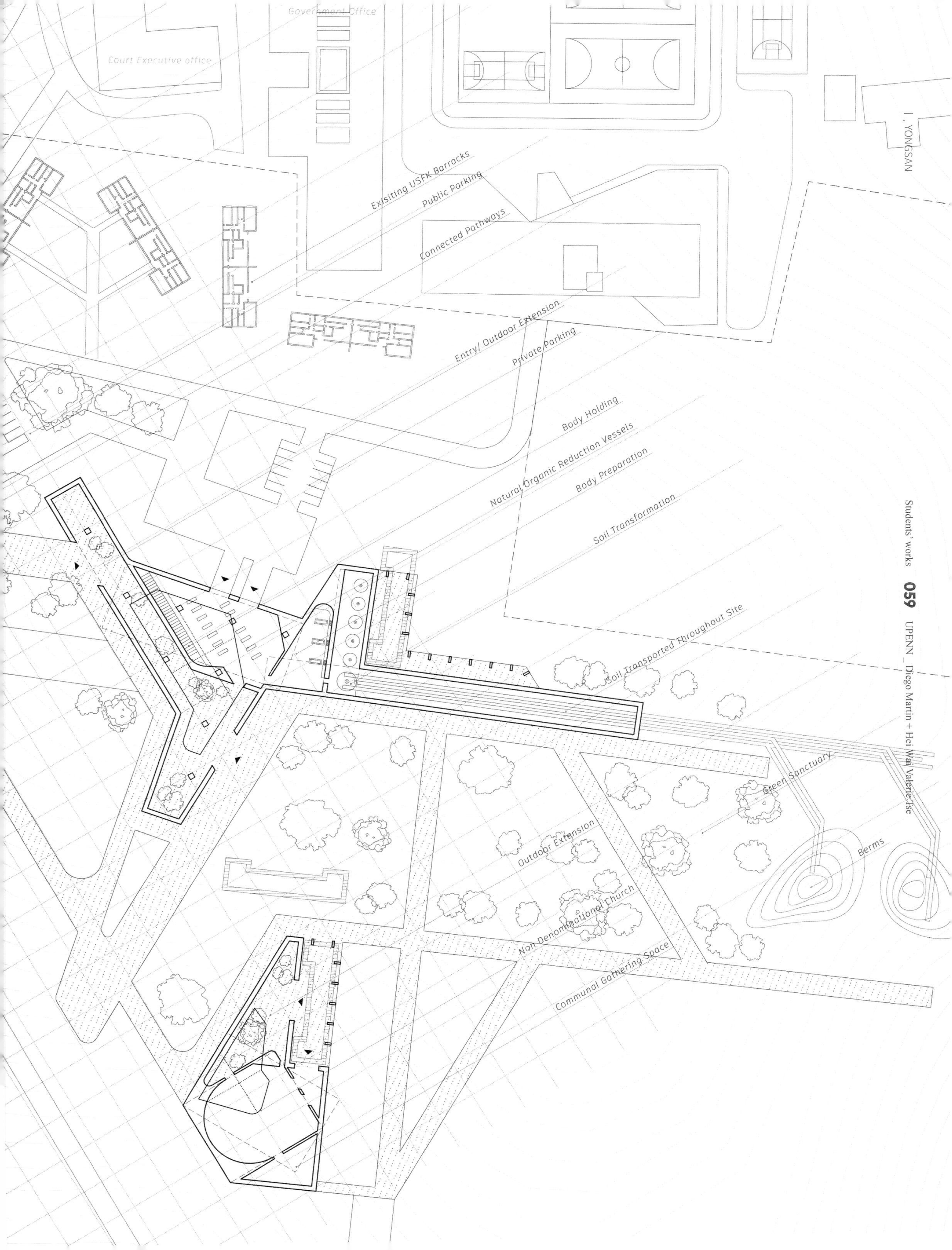
Government Office
Court Executive office
Exisiting USFK Barracks
Public Parking
Connected Pathways
Entry/ Outdoor Extension
Private Parking
Body Holding
Natural Organic Reduction Vessels
Body Preparation
Soil Transformation
Soil Transported Throughout Site
Green Sanctuary
Berms
Outdoor Extension
Non Denominational Church
Communal Gathering Space

SEED
EXCHANGE

ARCHIVING ARCHIPELAGO:
Reprogrammed Center for Community Archival Practices

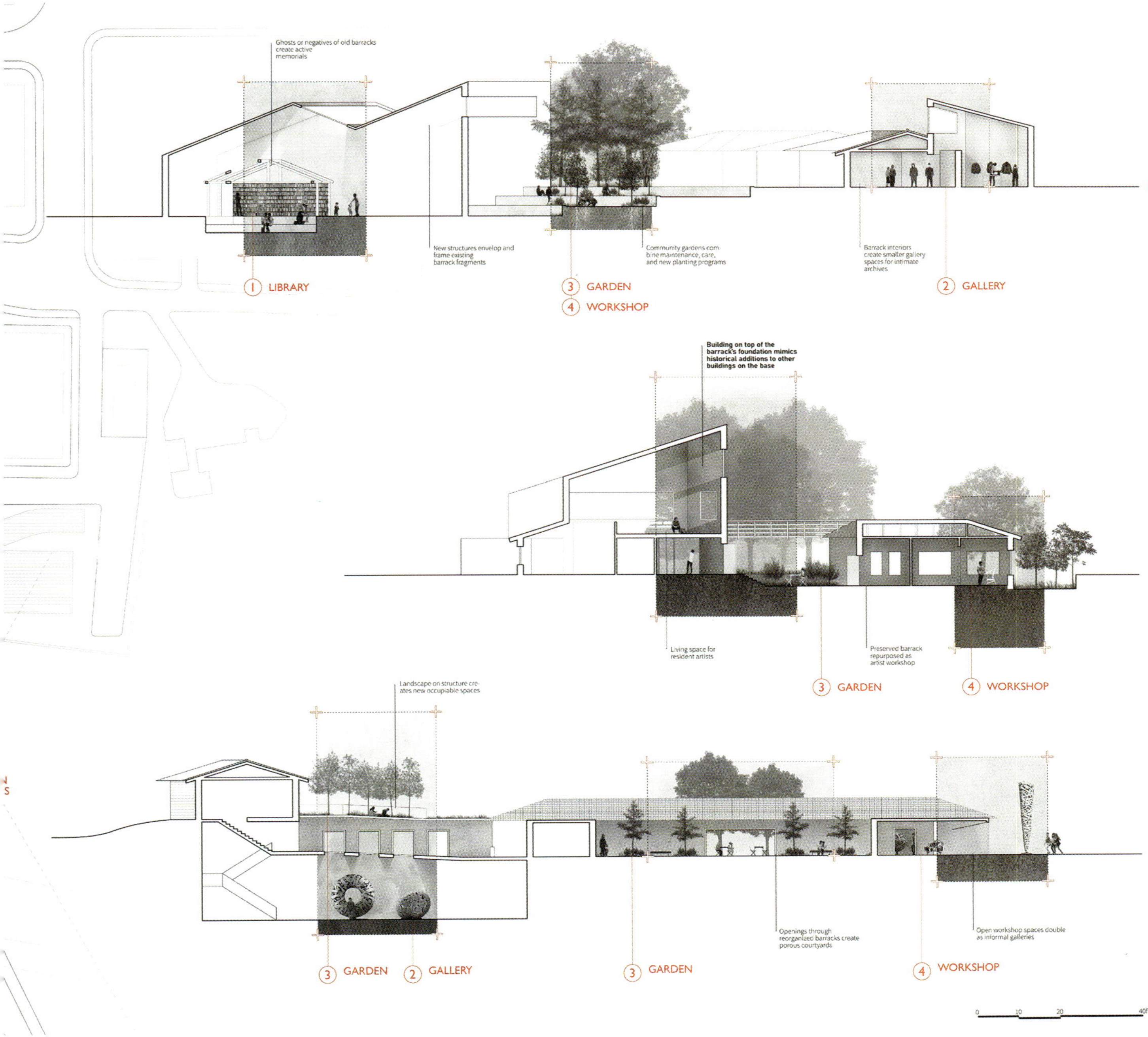
Ghosts or negatives of old barracks create active memorials
New structures envelop and frame existing barrack fragments
Community gardens combine maintenance, care, and new planting programs
Barrack interiors create smaller gallery spaces for intimate archives
A
1 LIBRARY
3 GARDEN
4 WORKSHOP
2 GALLERY
Building on top of the barrack's foundation mimics historical additions to other buildings on the base
Living space for resident artists
Preserved barrack repurposed as artist workshop
B
3 GARDEN
4 WORKSHOP
Landscape on structure creates new occupiable spaces
Openings through reorganized barracks create porous courtyards
Open workshop spaces double as informal galleries
C
3 GARDEN
2 GALLERY
3 GARDEN
4 WORKSHOP
0
10
20
40ft

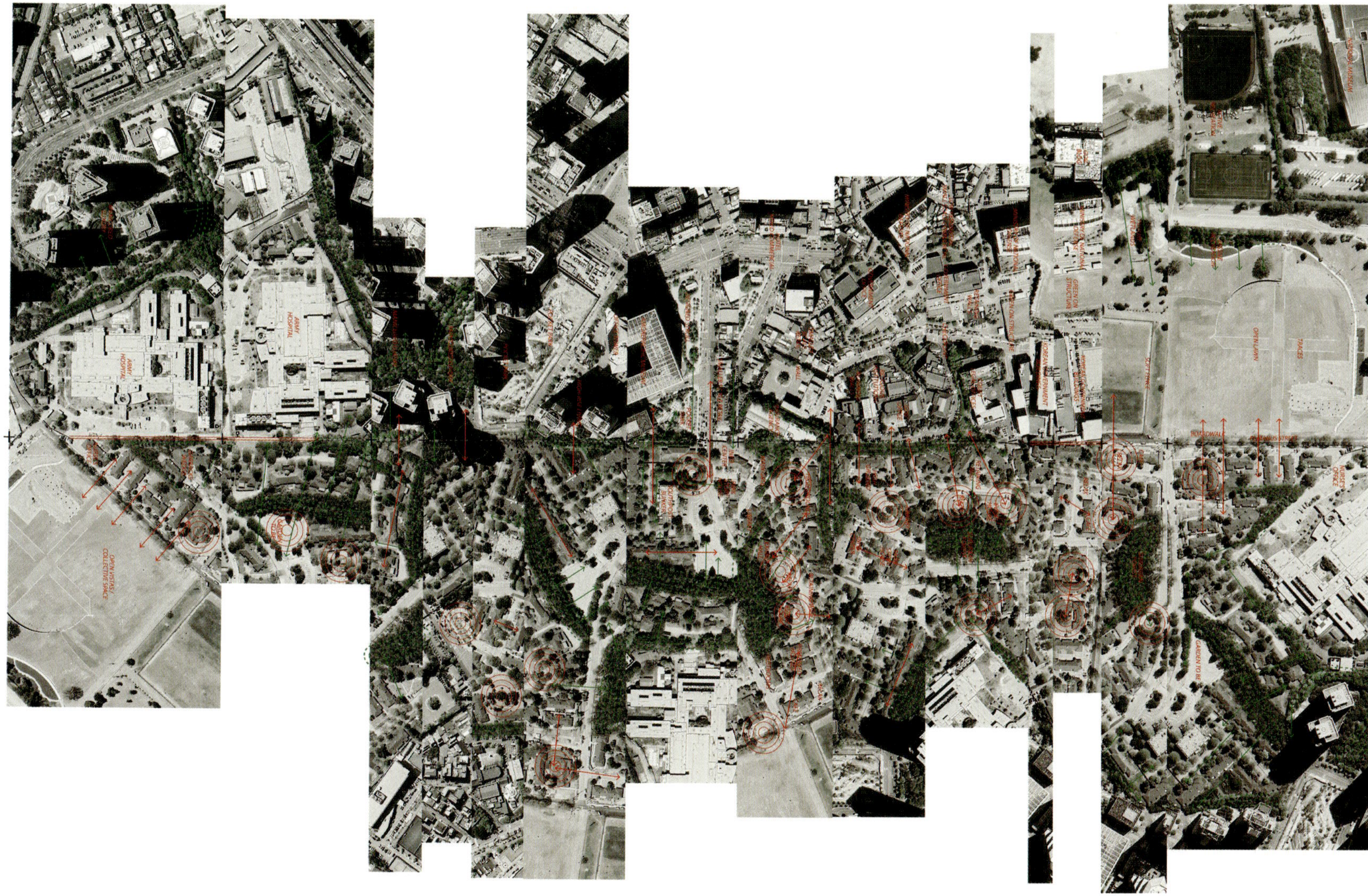
NATIONAL MUSEUM
TRACES
OPEN LAWN
NURSE'S OFFICE
GARDEN TO BE
SOFT FENCE
GREEN ON STRUCTURE
KOREAN GOVERNMENT
ARMY HOSPITAL
ARMY HOSPITAL
OPEN VISTAS / COLLECTIVE SPACE

This proposal addresses the decommissioning of the former Yongsan Garrison military base in Seoul, South Korea, as an architectural project and proposes creating the infrastructure for living archival practices. Treating layered narratives and material histories as opportunities for creative preservation and community storytelling, the reprogrammed base becomes a series of platforms for ongoing exchange across space, cultures, and generations.

As a critique of island logic and the totalizing narratives of the competitive urban park, this project focuses on domesticating the base's former barracks and grounds by hosting resident archival agents—the gardener, preservationist, and artist—to lead the community through ongoing practices of care and cultivation, deconstruction and renovation, and the creation of new artworks. With spaces that include a planting center, a preservationist gallery and workshop, an archive center, and artist studios, as well as public outdoor space, the reprogrammed base recognizes the stories that already exist on the site and the storytelling that it can continue to facilitate.

In this way, the project is a prototype for what can be done in other parts of Yongsan Garrison and with other bases so that the process of decommissioning is architectural, participatory, and focused on collective practice.

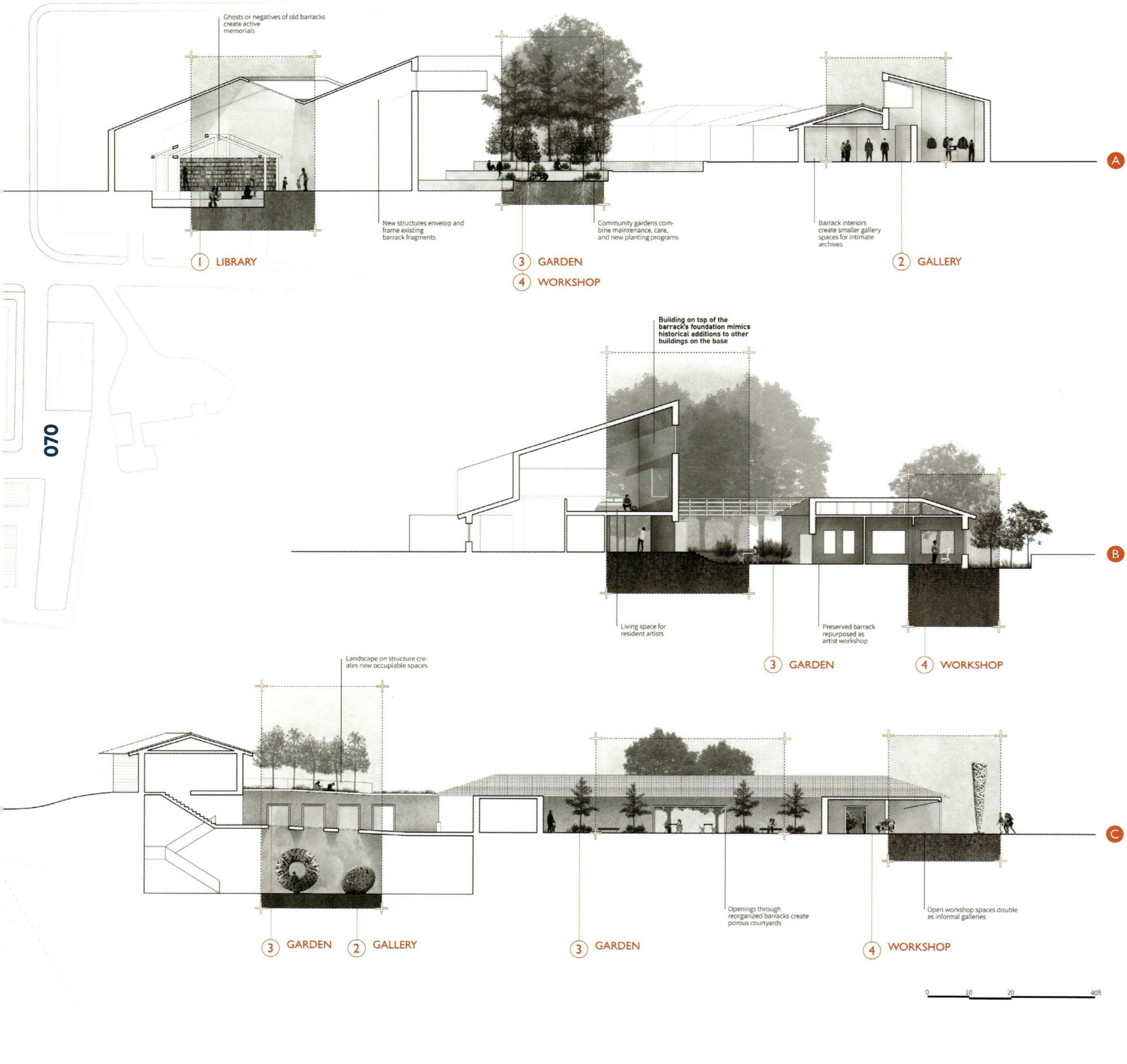
Ghosts or negatives of old barracks create active memorials
New structures envelop and frame existing barrack fragments
Community gardens combine maintenance, care, and new planting programs
Barrack interiors create smaller gallery spaces for intimate archives
A
1 LIBRARY
3 GARDEN
4 WORKSHOP
2 GALLERY
Building on top of the barrack's foundation mimics historical additions to other buildings on the base
B
Living space for resident artists
Preserved barrack repurposed as artist workshop
3 GARDEN
4 WORKSHOP
Landscape on structure creates new occupiable spaces
C
Openings through reorganized barracks create porous courtyards
Open workshop spaces double as informal galleries
3 GARDEN
2 GALLERY
3 GARDEN
4 WORKSHOP
0
10
20
40ft

Cannon Sculpture
Nam-Dan Stone Blocks
Urban: Light Industrial
Suburban: Group Barracks
Japanese Bachelor Officer Quarters
Ordinance Depot & Troop Barracks
Urban: Industrial
White House
War Memorial of Korea
8th Army Monument
Japanese Army Infantry Regiment Barracks
Suburban: Family Housing
Landscape: Interior Forest
Japanese Bachelor Officer Quarters
President's Office
Japanese Military Stockade
Dragon Hill Lodge
Japanese Housing
Suburban: Family Housing
Ministry of National Defense
Government Office
Japanese Bachelor Officer Quarters
Suburban: Unaccompanied Housing
South Post Bunker
Air Raid Shelter
Army Community Hospital
National Museum of Korea
Landscape: Meadow
Landscape: Shrubby Wetland
Landscape: Cleared Lawn
Suburban: Village

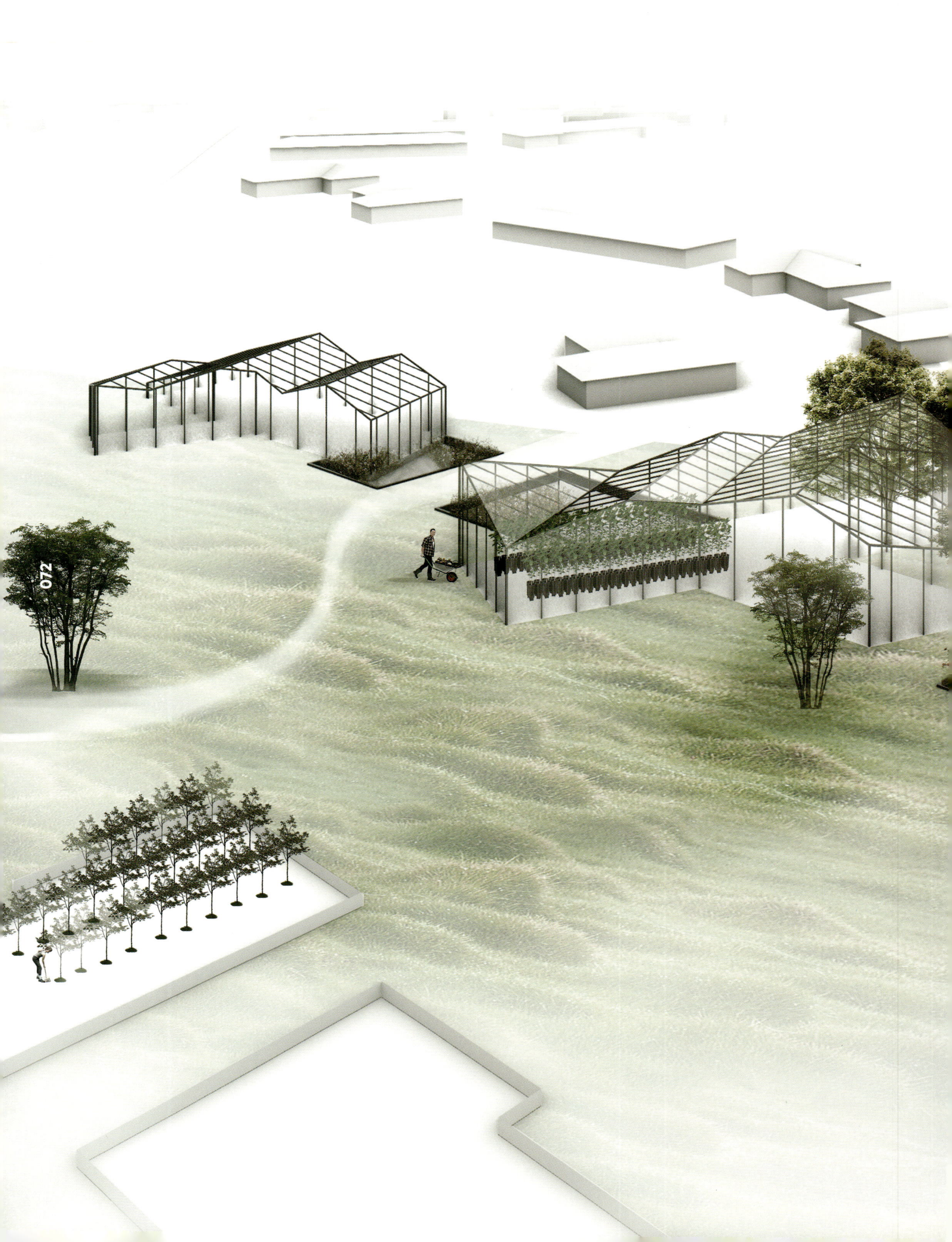

THE REPROGRAMMABLE CITY, SEOUL

TERRA INCOGNITA

SONG HYEON DONG

TERRA INCOGNITA
SONG
HYEON
DONG

II

SONG HYEON-DONG

~~Seoul Urbanism~~ Songhyeon Urbanism

~~Retroactive Reading of the Urban Form~~
Projecting towards the Terra Incognita

~~Sangheon Lee~~ Shinwoo Park, Yeongchae Kim, Yoojin Jeong

Fisrt Published in ~~November 2022~~ October 2023

1. ~~서울의 주요 경관요소는 결국 자연과 도시가 포개진 상황으로부터 나온다.~~[1]
2. 자연과 도시를 섬세하게 이으며 균형을 도모하는 태도가 '서울 어바니즘' 아닐까?
3. 그렇다면, 송현동은 서울 어바니즘을 실험해볼 수 있는 가장 알맞은 테스트베드다.
4. 서울을 구성하는 3개의 인지적 바닥면을 산(위), 도시(중간), 물(아래)라고 해보자.[2]
5. 송현은 이들이 모두 접속되는 접면이다; 서울의 구성원리와 현황이 한 데 집약된다.
6. 송현 일대의 형태적 윤곽은 북악산에서 청계천으로 흘러가는 개천 방향에 근거한다.
7. 개천 주위로 오랜 시간 쌓인 도시조직이 근대적 격자와 길 하나를 두고 맞닿는다.
8. ~~이것은 시간, 프로그램, 환경적인 측면에서 비동시적인 것들의 동시적 충돌이다.~~
9. '충돌'을 설계언어로 환원해 송현에 새로운 해석을 덧입히는 고민이 필요하다.
10. 결국 이는 서울 어바니즘에 대한 고민이다; ~~좋든 싫든 서울은 이질적이고 혼성적이다.~~[3]

(1) 피터 W. 페레토, 『플레이스/서울』, 프로파간다(2015)

(2) 이상헌, 『서울 어바니즘』, 공간서가(2022)

(3) 김성홍, 『서울해법』, 현암사(2020)

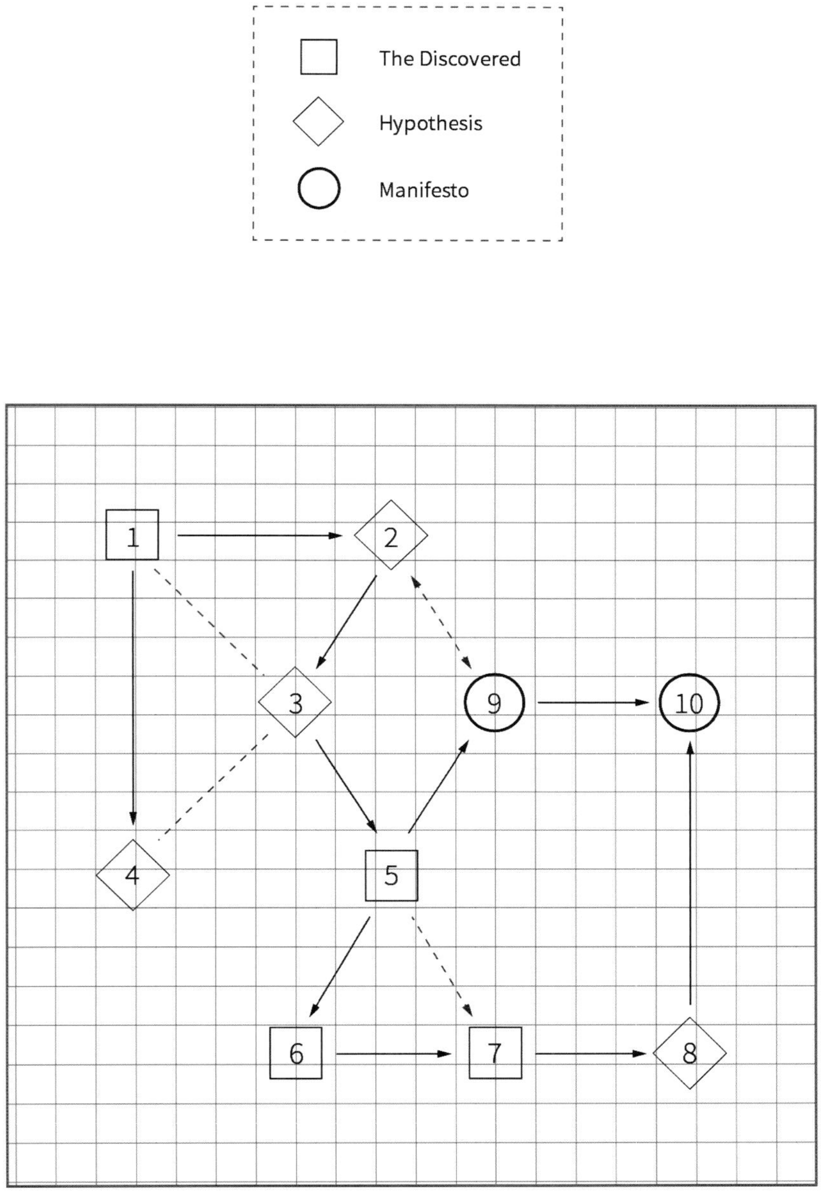
The Discovered
Hypothesis
Manifesto
1
2
3
4
5
6
7
8
9
10

Songhyeon Delaminated

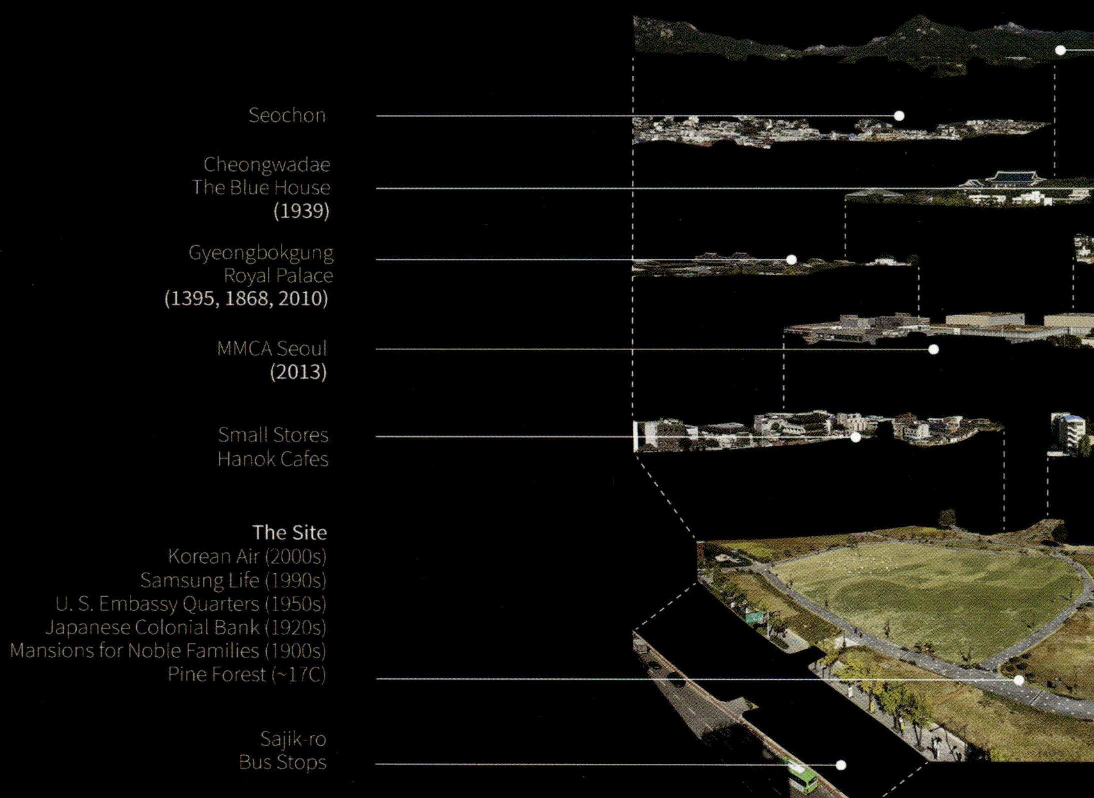

Image by Shinwoo Park

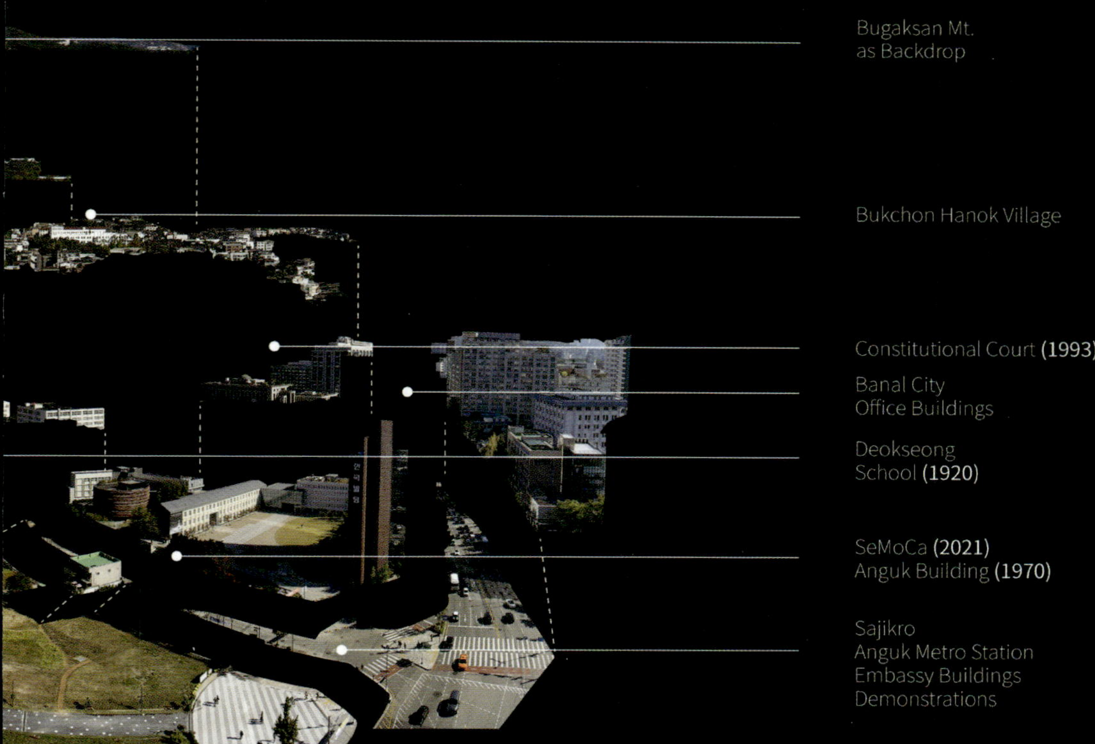
Bugaksan Mt.
as Backdrop
Bukchon Hanok Village
Constitutional Court (1993)
Banal City
Office Buildings
Deokseong
School (1920)
SeMoCa (2021)
Anguk Building (1970)
Sajikro
Anguk Metro Station
Embassy Buildings
Demonstrations

Songhyeon History

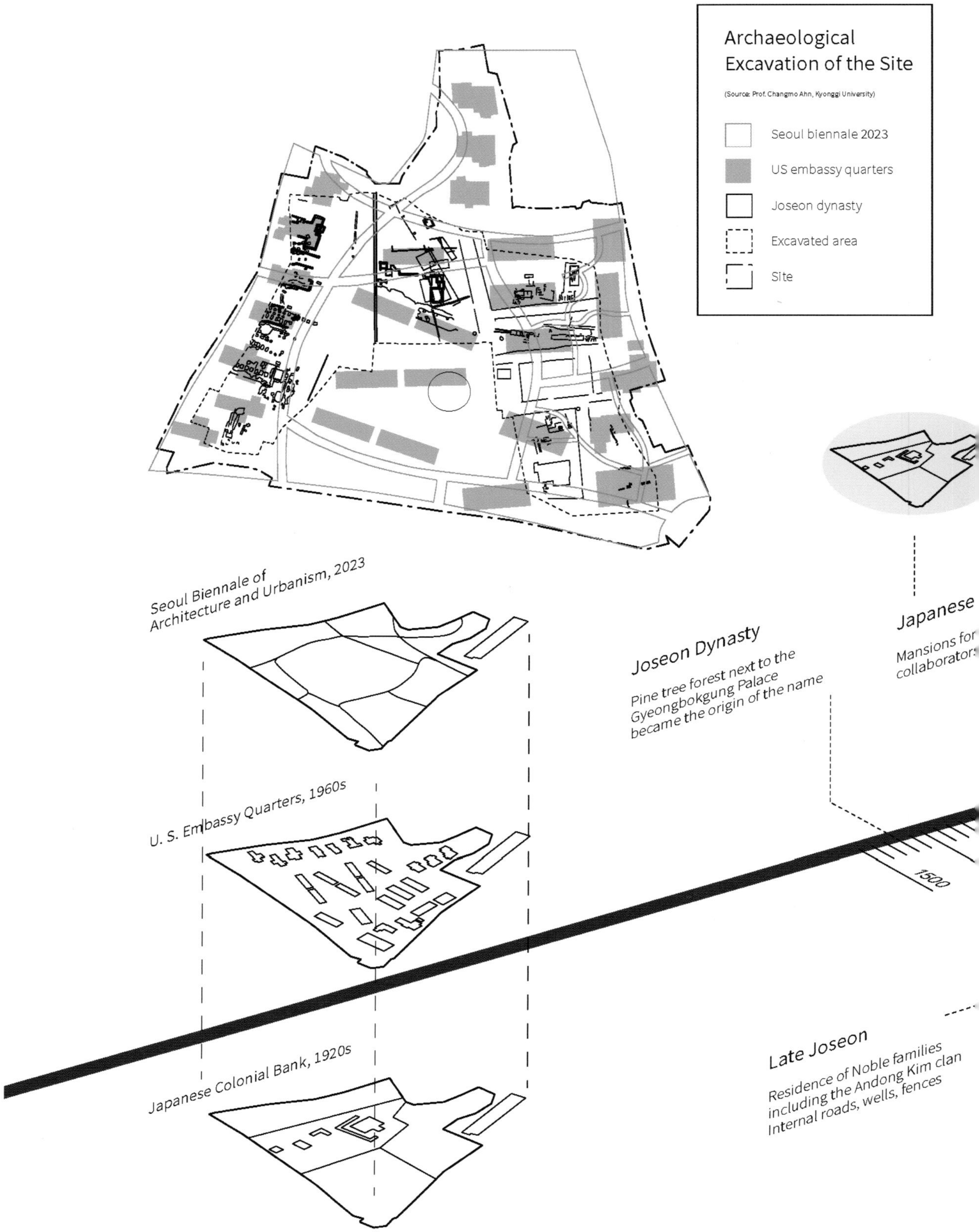

Image by Shinwoo Park

al Period

n
nily

Japanese Colonial Period

Company house for the
Japanese colonial bank (Siksan Bank)

Post - World War II

Quarters of
the United States embassy

IMF Crisis in Korea

Samsung planned an art museum,
failed due to the economical crisis

2000s, Korean Air

planned a hotel, failed due to the educational law
which bans the hotel near the school

2020s, Seoul Metropolitan Government

acquired the site from Korean Air,
utilizing it as a public park for citizens &
Seoul Biennale of Architecture and Urbanism 2023

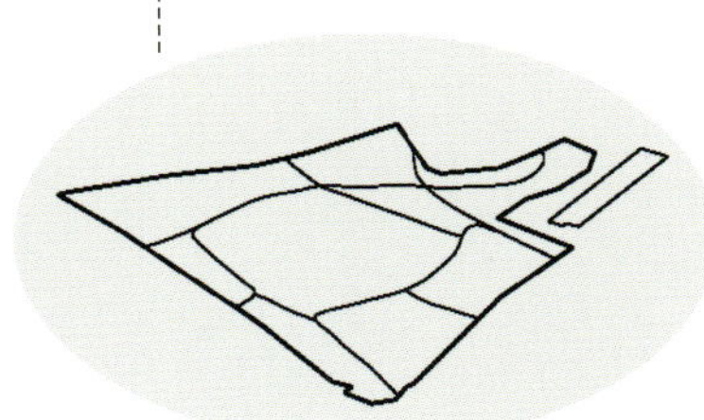

2030s

Seoul metropolitan government
planning new art museum

1700 1800 1900 1920 1940 1960 1980 2000 2020 2040

Art and Gallery

Tourism

Building Height

Program

Under ground

Hydrology

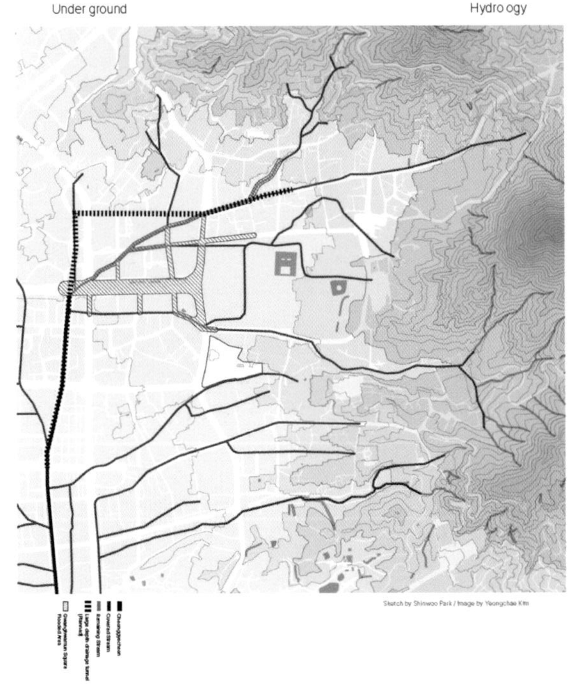

Under ground

Infrastructure

Eco ogy

Green

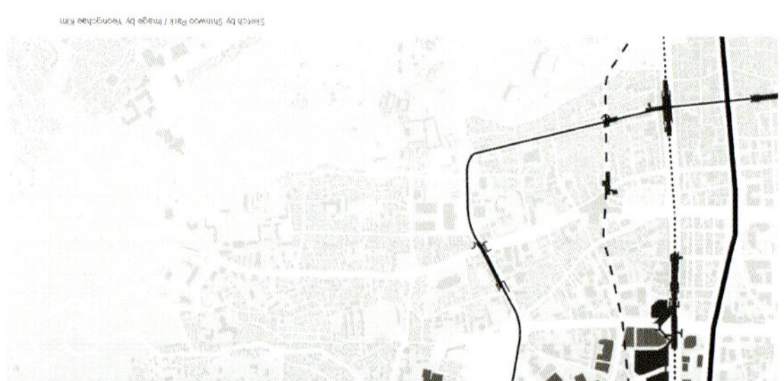

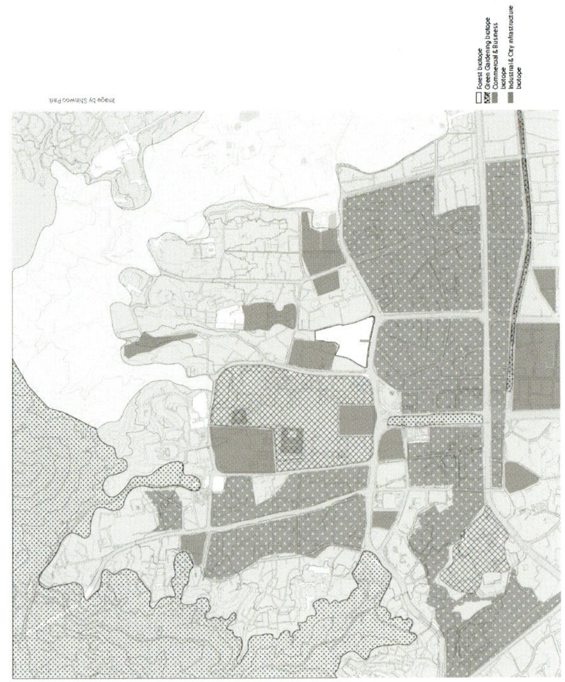

Eco ogy

Biotop

No I Map

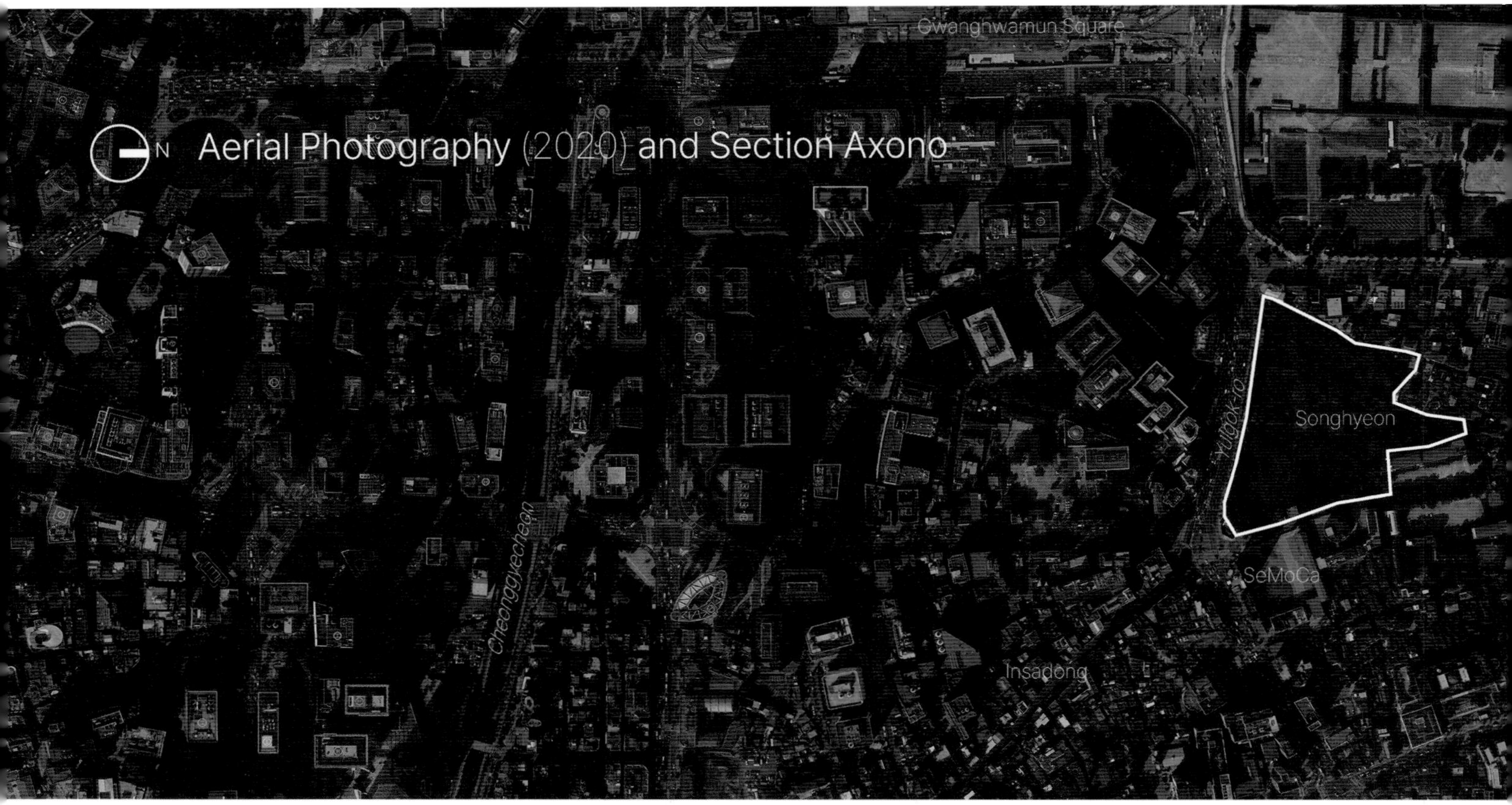
N
Aerial Photography (2020) and Section Axono
Gwanghwamun Square
Songhyeon
Yulgok-ro
Cheonggyecheon
SeMoCA
Insadong

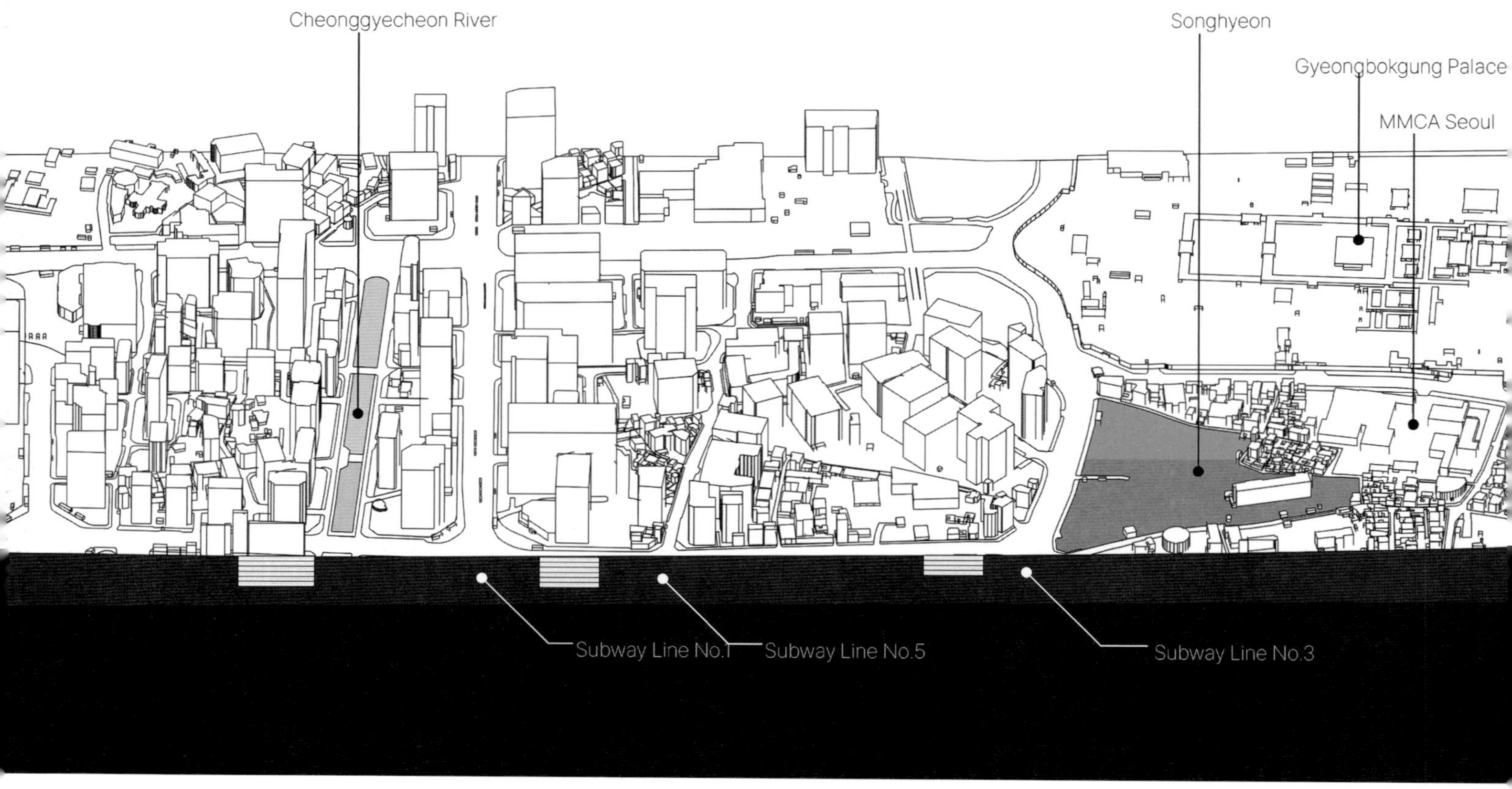
Cheonggyecheon River
Songhyeon
Gyeongbokgung Palace
MMCA Seoul
Subway Line No.1
Subway Line No.5
Subway Line No.3

Gyeongbokgung Palace
Bukaksan Mt.
Bukchon Hanok
Village

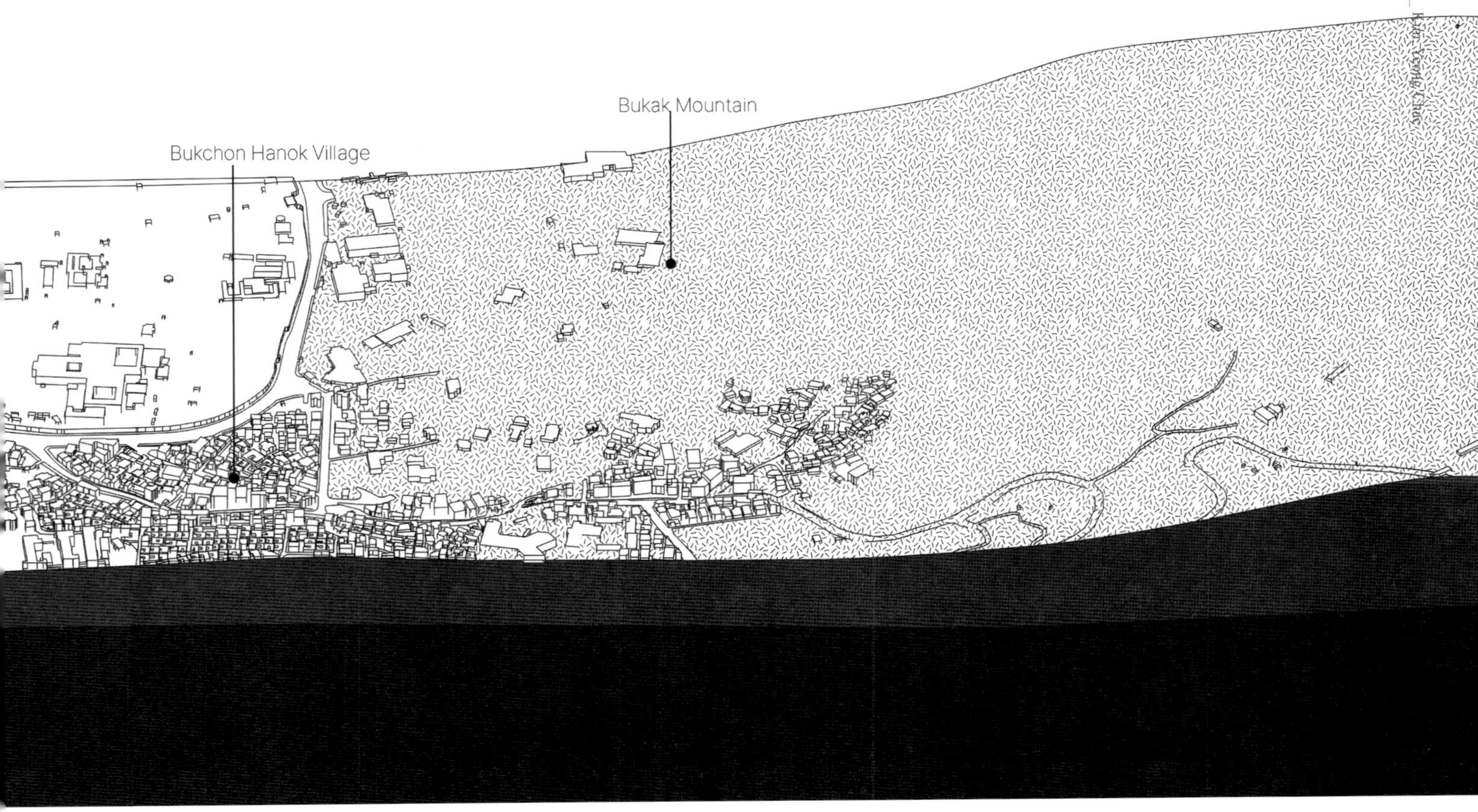
Bukak Mountain
Bukchon Hanok Village

Seoul Topography

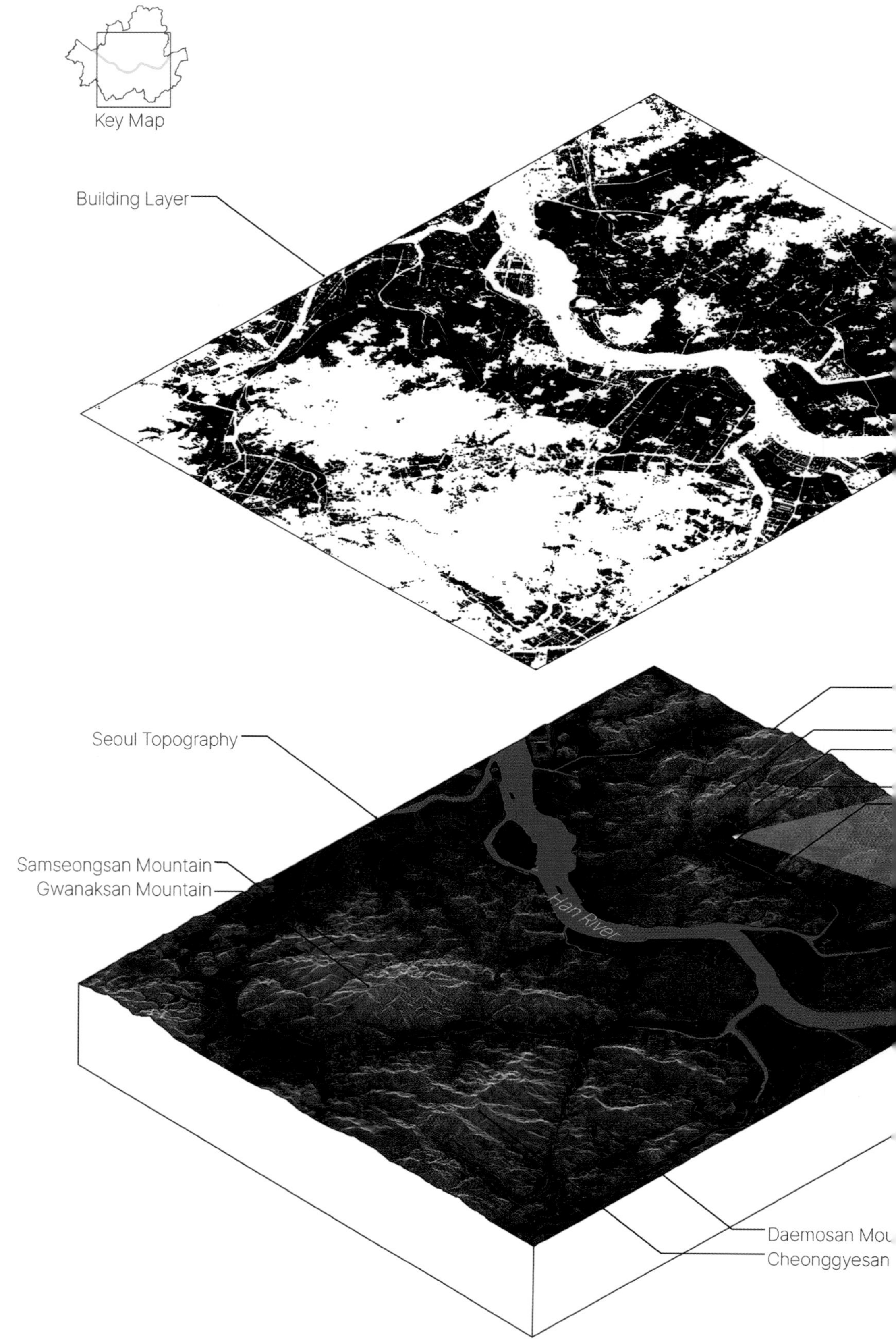

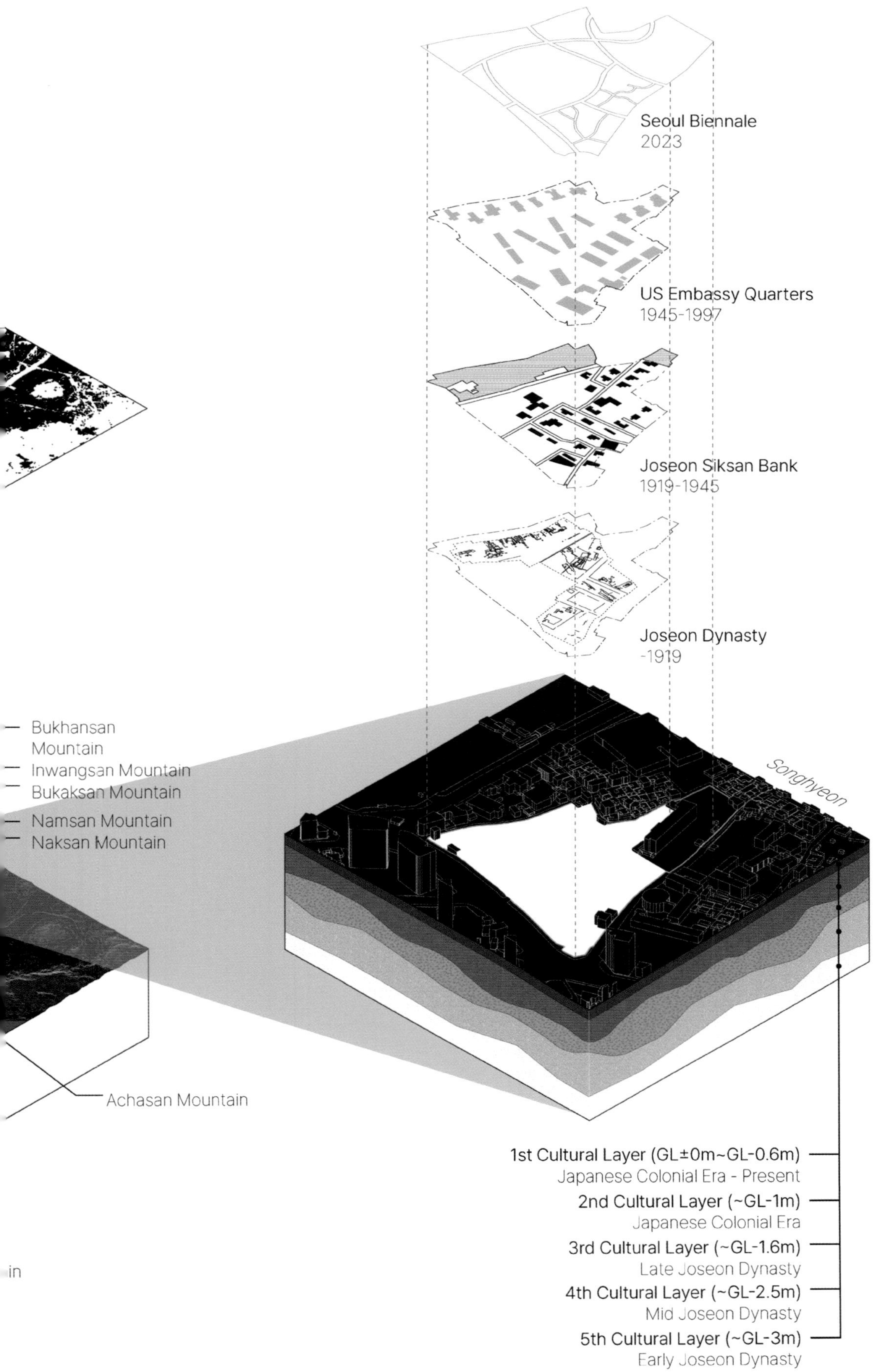
Seoul Biennale
2023
US Embassy Quarters
1945-1997
Joseon Siksan Bank
1919-1945
Joseon Dynasty
-1919
Bukhansan
Mountain
Inwangsan Mountain
Bukaksan Mountain
Namsan Mountain
Naksan Mountain
Songhyeon
Achasan Mountain
in
1st Cultural Layer (GL±0m~GL-0.6m)
Japanese Colonial Era - Present
2nd Cultural Layer (~GL-1m)
Japanese Colonial Era
3rd Cultural Layer (~GL-1.6m)
Late Joseon Dynasty
4th Cultural Layer (~GL-2.5m)
Mid Joseon Dynasty
5th Cultural Layer (~GL-3m)
Early Joseon Dynasty

N
SITE PLAN (GL-3200)
A3 1/2000
MMCA
Deokseong
Girls' High School
Deokseong
Girls' Middle School
SeMoCa
Yulgok-ro

B2F PLAN (GL-3200)
A3 1/400

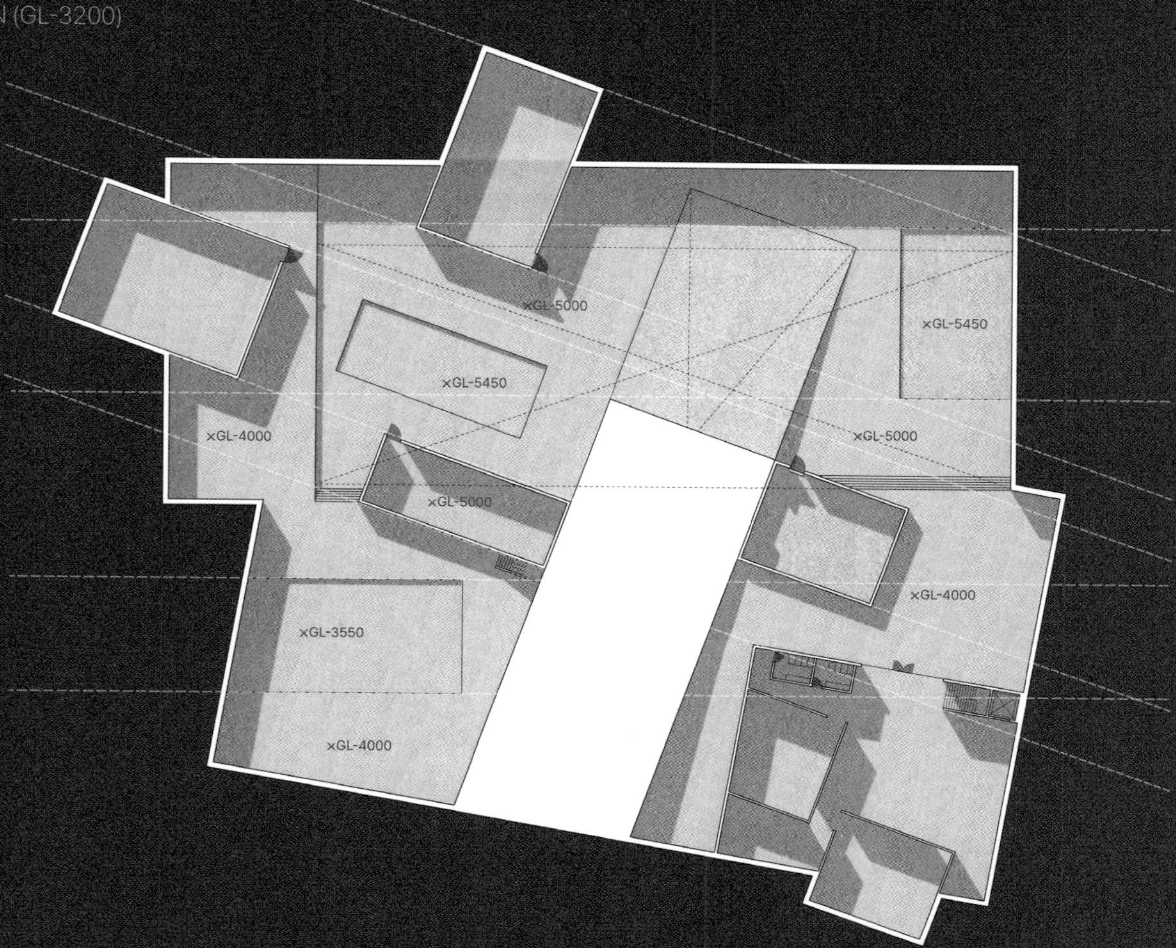

2F PLAN
A3 1/400

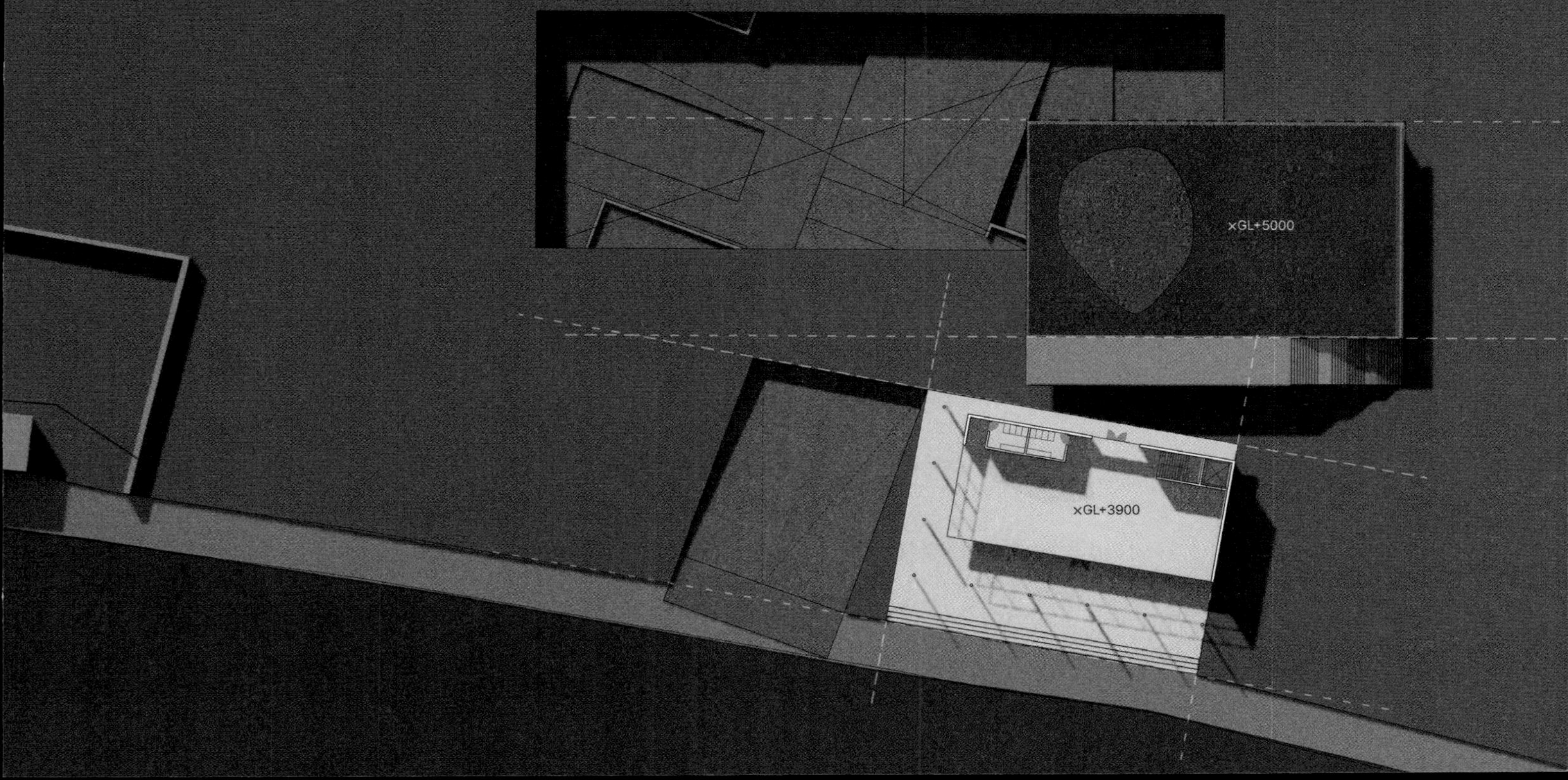

송현의 음각과 양각

서울은 "알프스 속의 맨해튼"이다. - 렘 콜하스

서울에는 외사산과 내사산을 비롯한 많은 산이 위치하고 자연적인 물길로 인해 도시조직의 형태가 불규칙하다. 시간차에 의한, 역사의 흔적이 남는 정도에 기인하는 서로 다른 필지의 모양과 크기는 곧 현재 서있는 건물의 스케일, 지하의 개발 정도의 차이로 이어진다. 이 두가지 요인은 서울이 입체적인 도시가 될 수밖에 없게 만들었다.
송현은 율곡로와 북악산자락의 사이에 위치한다. 송현의 주변으로는 서로 다른 시기에 개발된 도시조직들이 파편적으로 달라붙어있다. 이러한 차이는 주로 건물의 높이와 필지의 크기로 나타난다. 율곡로 남쪽으로는 고층 오피스 건물들이 주로 자리하며 깊이 개발된 지하층을 가지며 높이와 깊이의 차이가 더 극명하게 대조되도록 한다. 율곡로 북쪽으로는 오래된 한옥들과 낮은 저층건물들이 자리한다. 결과적으로 인공적인 요인과 자연적인 요인이 함께 작용하여 z축 상의 대조를 더 크게 만들며 송현을 사이에 두고 입체적인 도시 조직의 중첩이 극명해진다.

송현은 시간과 자연에 의한 z축의 대비, 역사와 현재, 일상과 비일상의 충돌이 가장 잘 드러나는 장소이다. 송현에서 이러한 충돌을 다루는 것이 곧 서울 어바니즘을 논하는 것과 다르지 않다.
광화문 광장 주변의 높은 오피스 건물들 - 율곡로 - 송현 - 북촌 한옥마을 - 북악산 까지 이어지는 다양한 높이의 레이어를 단면section적으로 바라보고자 한다.

이 프로젝트는 송현을 둘러싼 다양한 z축의 레벨을 아카이빙하는 것을 목표로 한다. 역사 레이어와 시간의 차이를 보여주기 위해 땅에는 음각과 양각을 만든다. 이는 시간의 차이에 의한 z축 상의 대비가 가장 극명하게 드러나는 송현의 남쪽, 율곡로를 따라 위치한다.

송현의 지하에는 깊어질수록 오래된 문화층이 존재하며, 근대의 문화층일수록 어지럽게 뒤얽혀 있다. 메인 경사로는 사람들이 이러한 역사 레이어 사이를 따라 내려가도록 한다. 경사로 중간중간에는 어지럽게 뒤얽혀있는 판에 도달하며 이는 펜스 안에 박제된 죽은 유구로 존재하는 것이 아니라 도시에서 실제로 작동하는 공간이 된다. 단순히 그 역사를 설명하고 존중하기만 하는 의미에서 벗어나 사람들의 일상과 함께 뒤섞이도록 한다. 그렇게 송현의 땅을 드러낸다. 파낸 흙은 다시 지상에 쌓인다. 쌓인 구조물은 사람들을 일상의 시선에서 벗어나게 한다. 위에서 음각을 내려다보며 땅이 겪어온 수많은 흔적과 역사를 실감한다. 위로 올라갈 수록 멀리 바라보며 땅을 둘러싼 서로 다른 도시의 조각들을 겹쳐 본다.

Engraving and Embossing of Songhyeon

There are two main factors causing the extreme difference of heights in Seoul. First, Seoul has most dynamic topography. Next, because of the different development timeline of Seoul urban tissues, each part of the city has different physical features. So, the dramatic difference of building heights is made because of these two factors.
Songhyeon is located between Yulgok-ro and the starting of Bukak mountain. Moreover, Songhyeon is surrounded by different urban context, caused by the time and history. The difference usually appears by the heights and the scale of the surrounding elements. Below the Yulgok-ro, there are high rised office buildings. They usually have basement floors going deep down into the ground making the z-axis difference even more. Above the Yulgok-ro, in contrast, there are old, out-dated Hanoks with low-rised heights. As a result, environmental factor and artificial factor are acting together, maximizing the difference in z-axis and making Seoul even more 3-dimensional. And Songhyeon is located in the middle of the contrast.

The layering and contrast around Songhyeon are not much different from that of Seoul. So, I would like to view Songhyeon's layer as the huge section across Gwanghwamun Square to Bukak Mountain.

This project's goal is to archive the different z-axis around Songhyeon of different time line. To show the tracing of historical layers, I made the engraving and embossing on the southern part of Songhyeon where the contrast of the time and physical scale is most extreme.

There are cultural layers under the Songhyeon, which get older as it goes deeper. There are layers of Joseon Dynasty, layer of Korean Empire Age, layer of Japanese Colonial Era, and so on. Main ramp of the structure leads people to the layer of history underground. As people goes down, they arrive at different historical layer, which follows the tracing of historical layer. Each layer becomes a 'place' functioning within the city instead of being 'dead' inside the fence as preserved remains. In that way, this project reveals the Songhyeon's layer.

The earth that was digged out is re-accumulated on the ground of Songhyeon. The structure above ground allows people to go up and to have different view of sightseeing the city. They can enjoy the overlapped layer of engraving looking down the sunken. They can have wider view as they go up, allowing themselves to overlap each different part of Jongno gu from different historical urban background.

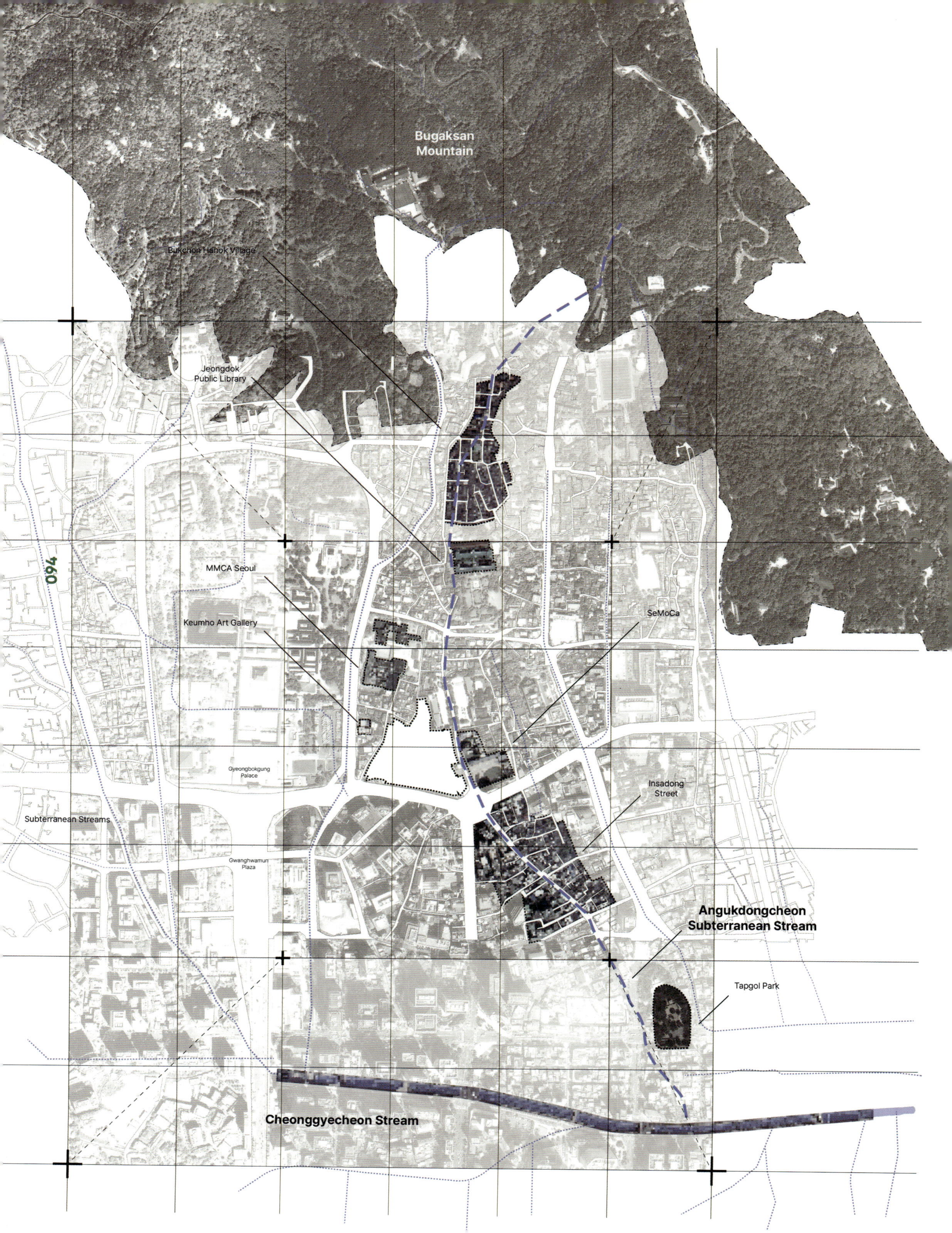
Bugaksan
Mountain
Bukchon Hanok Village
Jeongdok
Public Library
MMCA Seoul
Keumho Art Gallery
SeMoCa
Gyeongbokgung
Palace
Insadong
Street
Subterranean Streams
Gwanghwamun
Plaza
Angukdongcheon
Subterranean Stream
Tapgol Park
Cheonggyecheon Stream

물길

급작스럽게 개방된 송현동을 안정화하고 점진적으로 열어가는 전략으로서 필지 모서리에 면한 보행가로와 복개천에 초점을 맞춥니다. 닫혀 있던 이전적지를 길의 일부로 이해하는 이러한 시도는 대상지 남북으로 늘어선 도시경관에 문화적, 생태적 연속성을 더할 수 있습니다. 이는 산과 하천, 그리고 길을 기초로 하는 종로 일대의 공간구조와 맥을 같이하며, 마스터플랜으로 큰 땅을 장악하는 과거의 태도와 거리를 둡니다.

대상지 동쪽으로 면한 감고당길과 그 아래에 묻힌 안국동천은 재맥락화의 핵심 논거가 됩니다. 서울형 수변감성도시 계획에 근거한 안국동천의 복원은 인사동과 북촌을 잇는 산뜻한 보행환경을 조성하며 미기후 조절에 기여합니다. 특히 높은 건폐율과 배수계획의 미비함으로 투수율이 나쁜 북촌에서 여름철 홍수 피해를 일으키는 국지성 호우의 일정량을 담지합니다.

뿐만 아니라 안국동천의 복원은 지하수를 매개로 희미하게 연결된 북악산과 청계천의 생태환경을 한층 결속하고 가시화할 수 있습니다. 이에 따라 대상지 주변의 비오톱과 야생종 분포, 계절별 강수량 데이터를 바탕으로 개천의 형태, 용량, 조경수종이 결정되었습니다.

수변을 따라 놓인 미술관 매스들과 인공의 지형은 계절에 따라 물의 통과를 허용하며 양쪽 입면이 모두 파사드가 될 수 있습니다. 다양한 단면의 매스들을 매개하는 공중의 회랑은 입면에 통일성을 부여하며 일반 시민과 관람객, 그리고 스태프의 동선을 정리합니다. 회랑은 폐교가 예상되는 대상지 북쪽의 덕성여중으로 관입되어 기존 건물의 코어와 연결되기도 합니다.

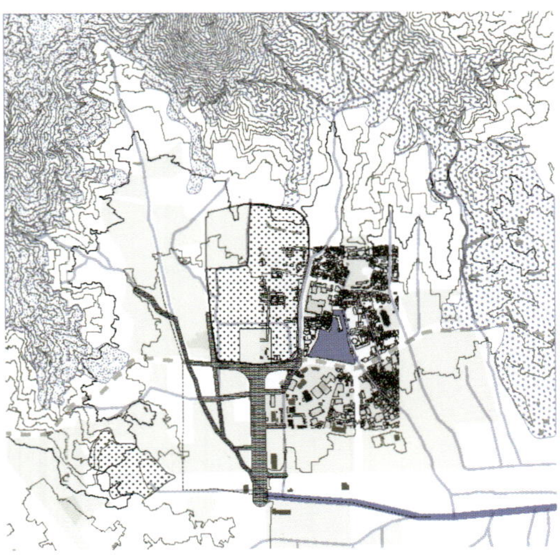

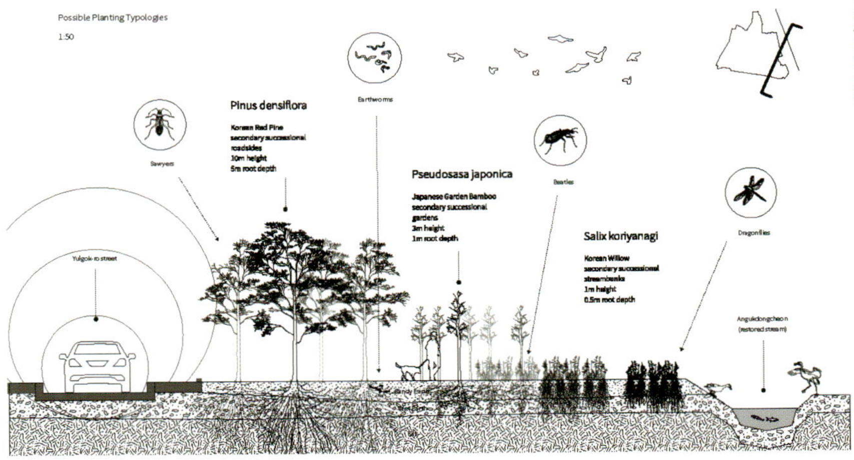

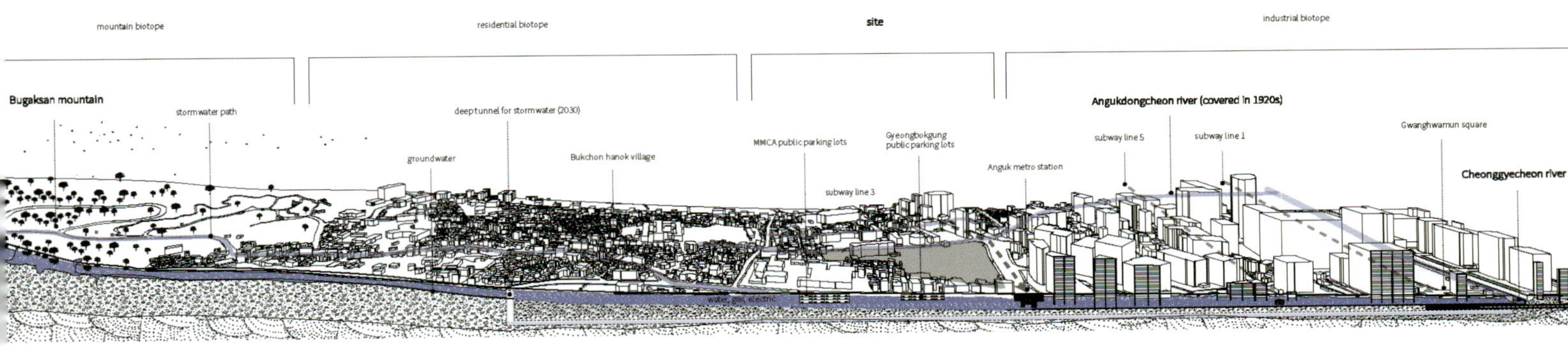

Waterway: daylighting stream as urban design strategy in Songhyeondong

To stabilize and gradually contextualize Songhyeon-dong, a former U. S. army base in the center of Seoul, this project focused on the pedestrian street and the covered stream facing the edge of the site. This understanding of the previously closed site as part of the street can add cultural and ecological continuity to the urban landscape. This aligns with the Jongno area's spatial structure based on mountains, water, and streets while distancing itself from past attitudes to overwhelm extensive lands with master plans.

The critical factors for re-contextualization are the Gamgodang-gil street facing the east of the site and the Angukdongcheon stream buried under it. Deculverting the Anguk-dongcheon stream based on the Seoul Waterfront City Plan creates a fresh walking environment linking Insa-dong and Bukchon, contributing to microclimate control. It also contains a certain amount of local heavy rain that causes flood damage in summer in Bukchon, where the permeability rate is poor due to the high BCR and the lack of a drainage system.

This daylighted stream can further bind and visualize the ecological environment of Bugaksan Mountain and Cheonggyecheon Stream, which are connected only through groundwater. As a result, the stream's shape, capacity, and landscape tree species were determined based on the distribution of biotopes, wild species, and seasonal precipitation data around the site.

Both sides of each mass along the waterfront can be a facade depending on the dry and rainy season conditions. The cloister connects and mediates each mass with various cross-sections, organizing the circulations of passers-by, museum visitors, and staff.
The corridor also penetrates and connects the existing core of Deokseong Girls' Middle School which is expected to close soon.

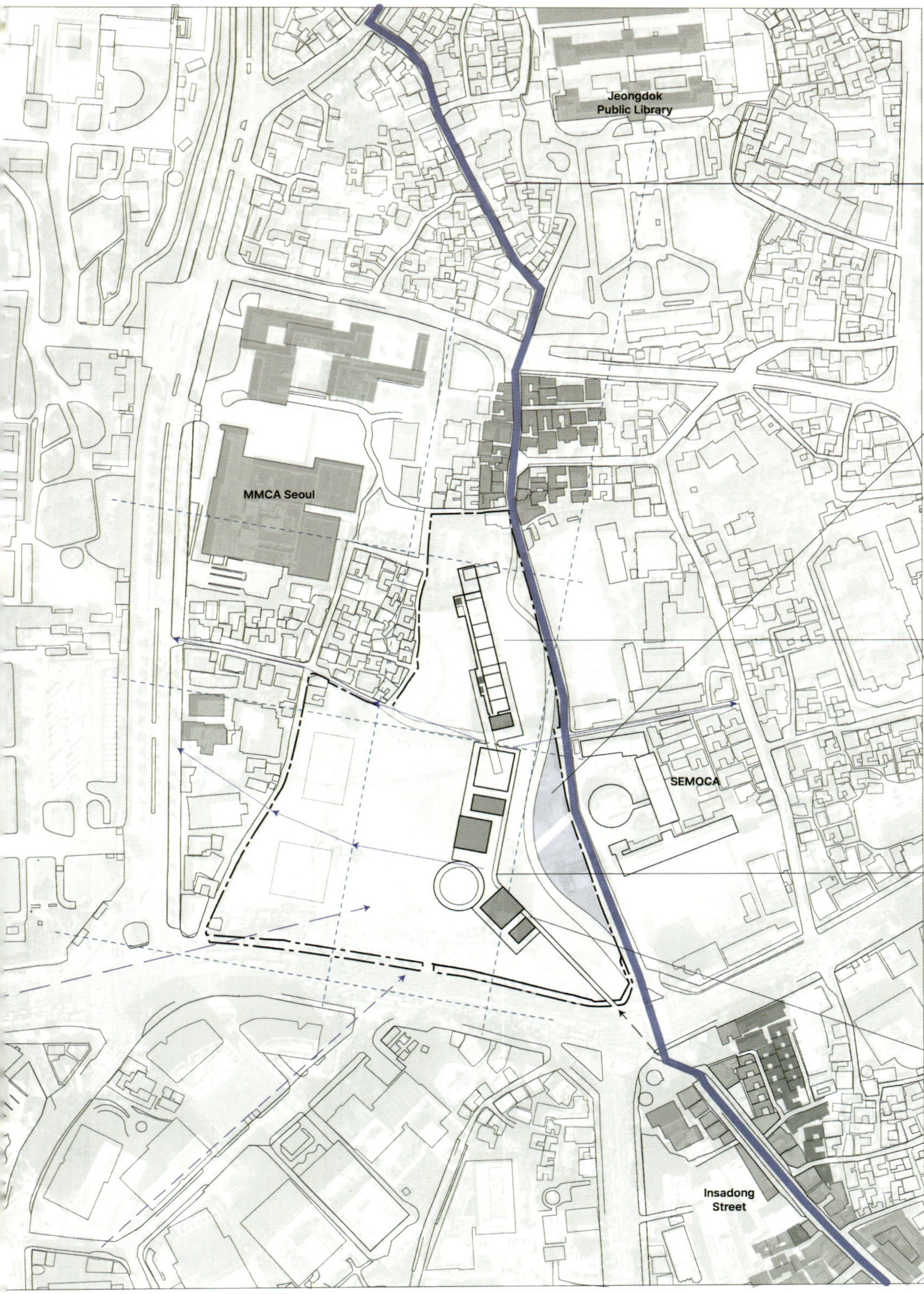

Impermeable Area

14,600m2 of Bukchon Hanok Village is impermeable due to high bldg coverage

Causing severe flooding every summer

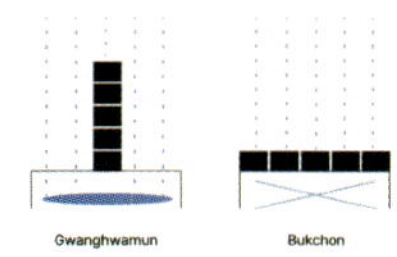

Water Basins

Restored from the subterranean streams
Basins hold 250m3 stormwater
Covers 180mm/day precipitation

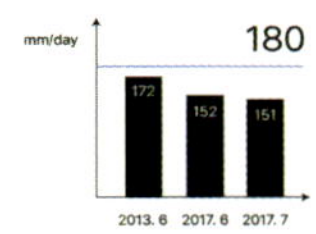

Deoksung School
Adaptable Reuse
Archive

Art Museum

Cloister
connects programs

Connecting the Existing School

Penetrates into the existing school which is about to close, due to population decrease. Stairs will be reused. Voids are made.

Expansion

Moving to the left, column spans narrow down little by little to adjust to the module of the existing school building.

Archive Type

Serves as a main circulation for the visitors while embracing the landscape inside the building. The landscape is partly contoured to be an exhibition stand. At the ground level is a visible storage of artworks. Archive and library on the upper floors connected by the escalator.

Curation Type

A main circulation for the staf

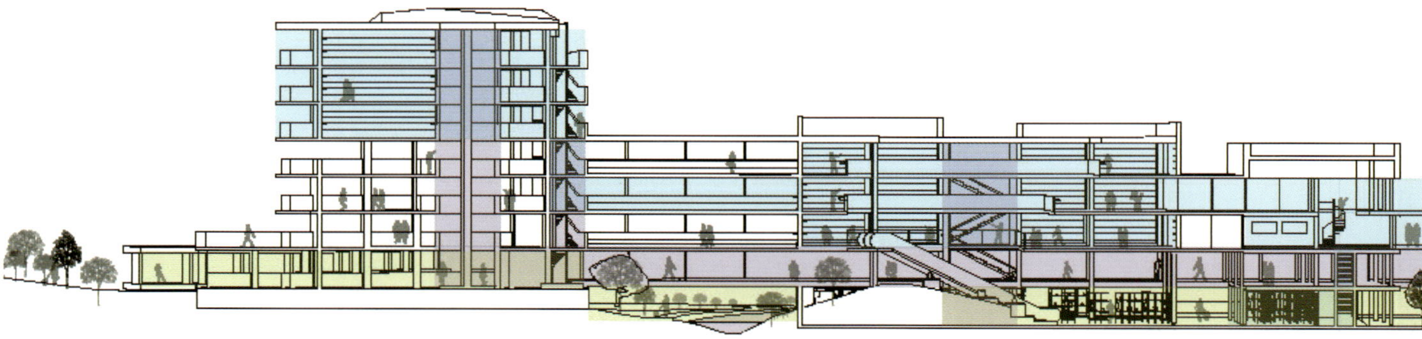

Secondary Water Basin

Serves as an urban corridor which catalyzes the east - west movement

Adaptive Reuse of the School

Adaptive Reuse of the School

Auditorium Type

Seats oriented towards the plaza on the west. The stage could be opened outside as shown at the individual section drawing. Functions as a theater or a multipurpose hall.

Spiral Type

Opens up a panorama view that embraces the waterfront. Cylinder shape corresponds to the SeMoCa museum across the street.

White Cube Type

Accomodates various scale of artworks from little craft arts to huge installations.

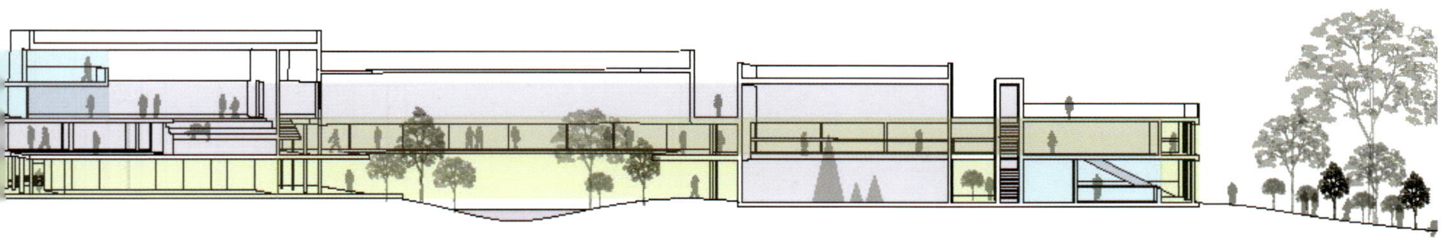

Primary Water Basin

Penetrates underneath the building when the stormwater floods at summer. Can be seen at the spiral gallery.

Staff

Museum Visitors

Passers-by

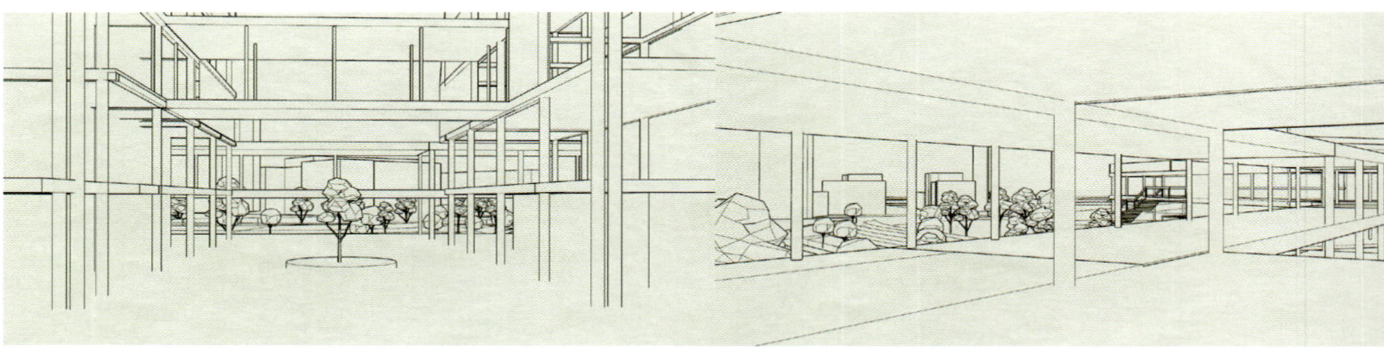

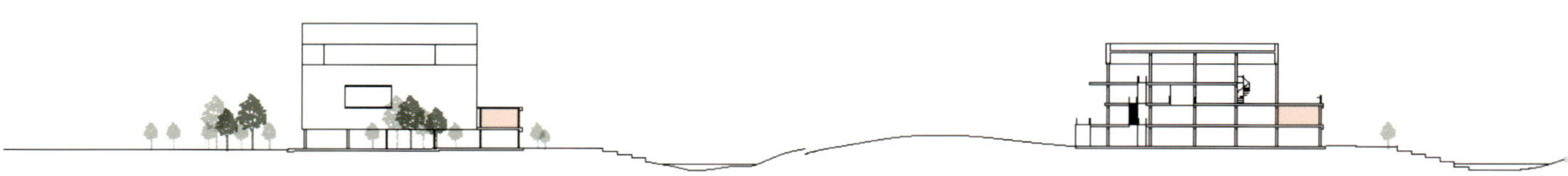

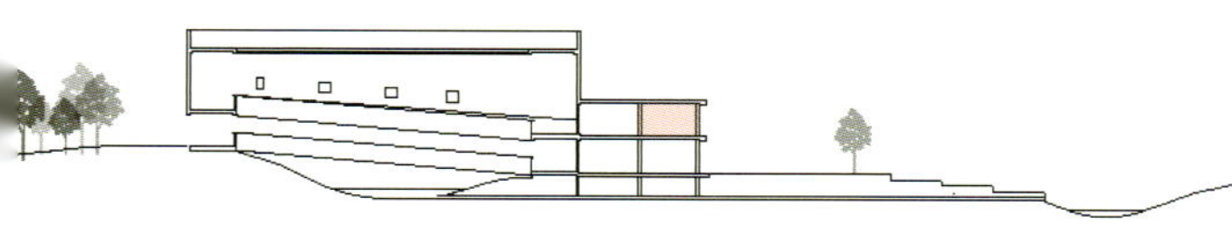

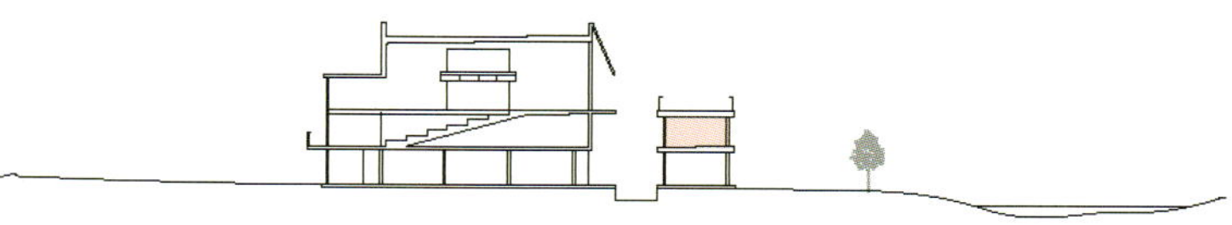

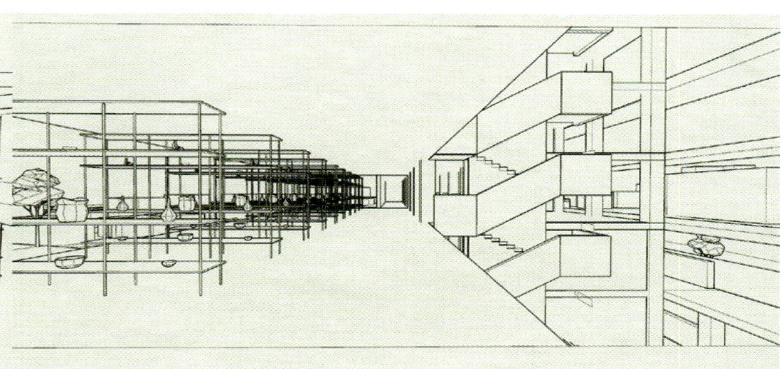

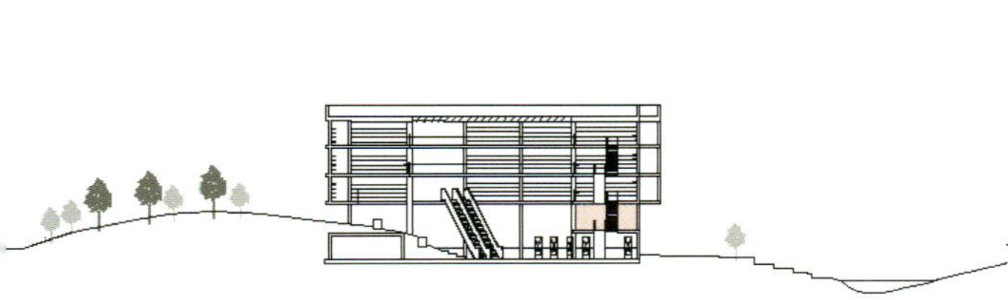

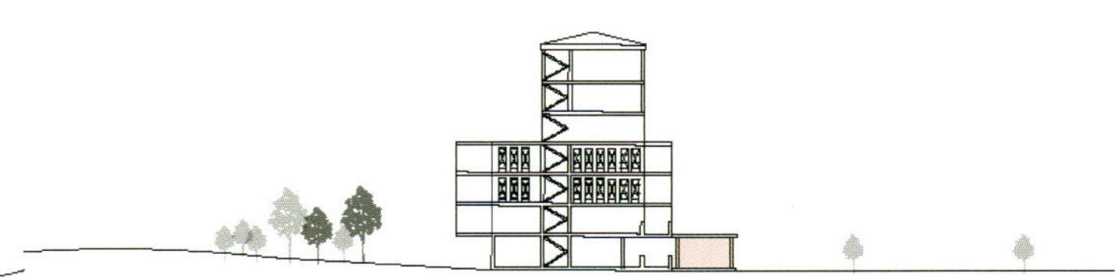

Bathscape: Weaving Collective Experiences

Bathhouses, functioning as a third space, possess a multifaceted significance that transcends their utilitarian purpose. Beyond being mere facilities for bodily cleansing, they assume a pivotal role as communal hubs, where societal norms of social stratification and racial distinctions are subdued. Within these spaces, individuals from diverse backgrounds converge, enabling an environment where socio-economic status and racial barriers lose their potency. This dissolution of barriers fosters an egalitarian ethos, allowing for the convergence of various social strata and ethnicities on neutral ground.

The bathhouse, in its essence, embodies a microcosm of society, where the act of communal bathing becomes a conduit for social interaction and discourse. Strangers, often separated by societal constructs, converge here, engaging in dialogues that transcend the limitations imposed by their external identities. Conversations unfold, opinions are shared, and connections form, facilitated by the shared vulnerability and intimacy inherent in the act of cleansing oneself. This shared vulnerability creates a unique platform for intermingling, enabling the convergence of disparate perspectives, experiences, and narratives.

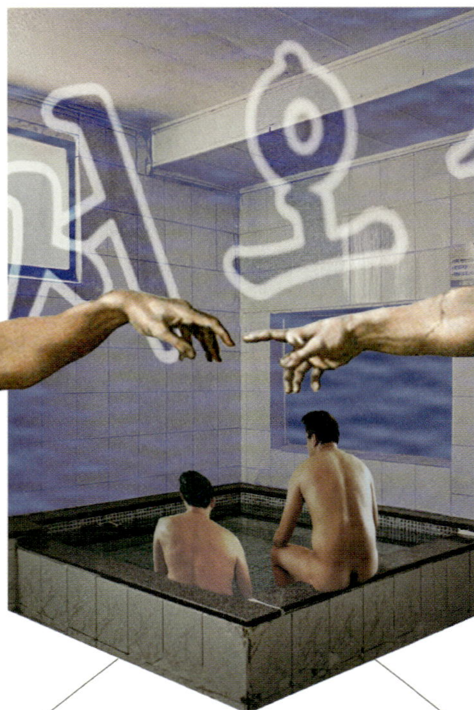

Brief History

Public baths emerged due to the scarcity of private bathing facilities in densely populated areas, serving a primary function of communal hygiene. Despite being labeled as "public," these spaces often enforced restrictions based on gender, religious affiliations, personal memberships, and other criteria. However, beyond their hygiene-oriented role, public baths evolved into significant social meeting points.

As societies witnessed an increase in households equipped with private bathrooms, the necessity for public baths declined, leading these spaces to predominantly serve recreational purposes. Notably, in Korean culture, public baths are renowned for their diverse amenities, extending beyond fundamental bathing facilities. These encompass a wide spectrum of offerings, such as public saunas, hot tubs, showers, and massage tables for patrons seeking scrub massages, elevating the experience beyond mere hygiene to encompass relaxation and wellness elements.

Scale of Communal baths

Hair Salons and Diversity

The Act of Bathing as Pure Rituals

Neighborhood Open Spaces Based on Characteristics

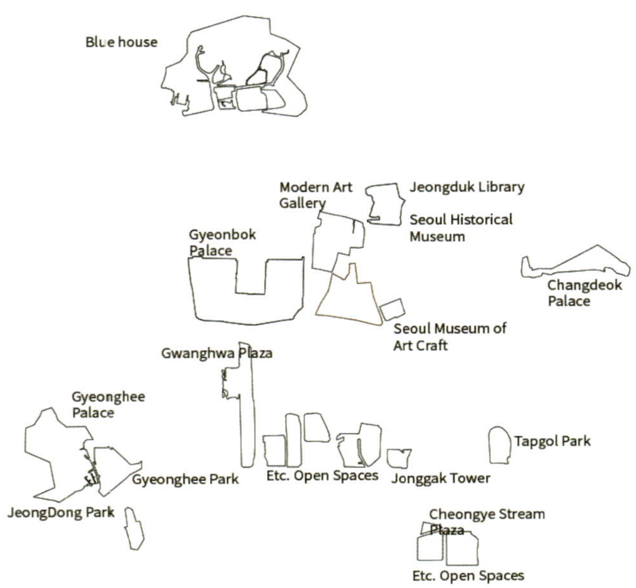

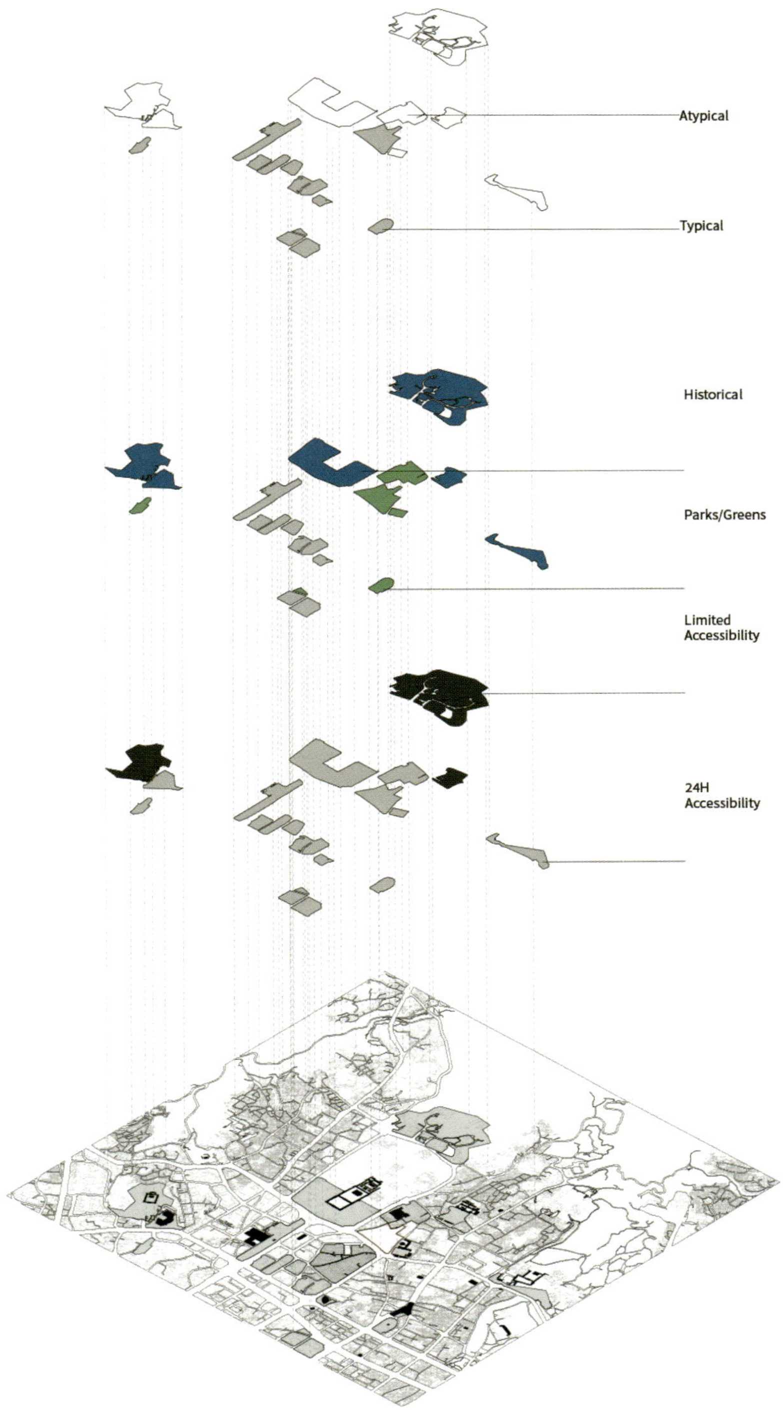

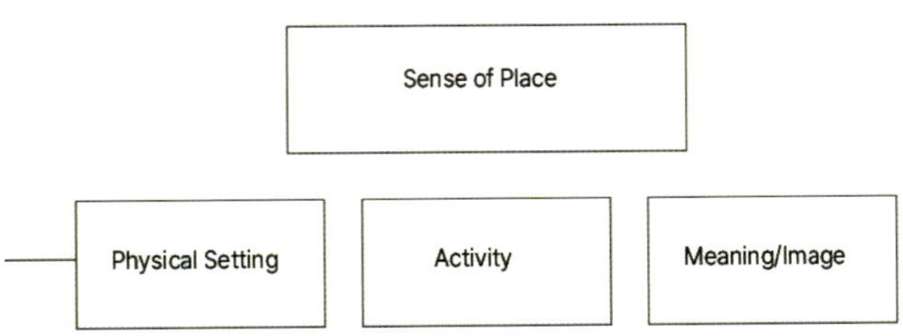

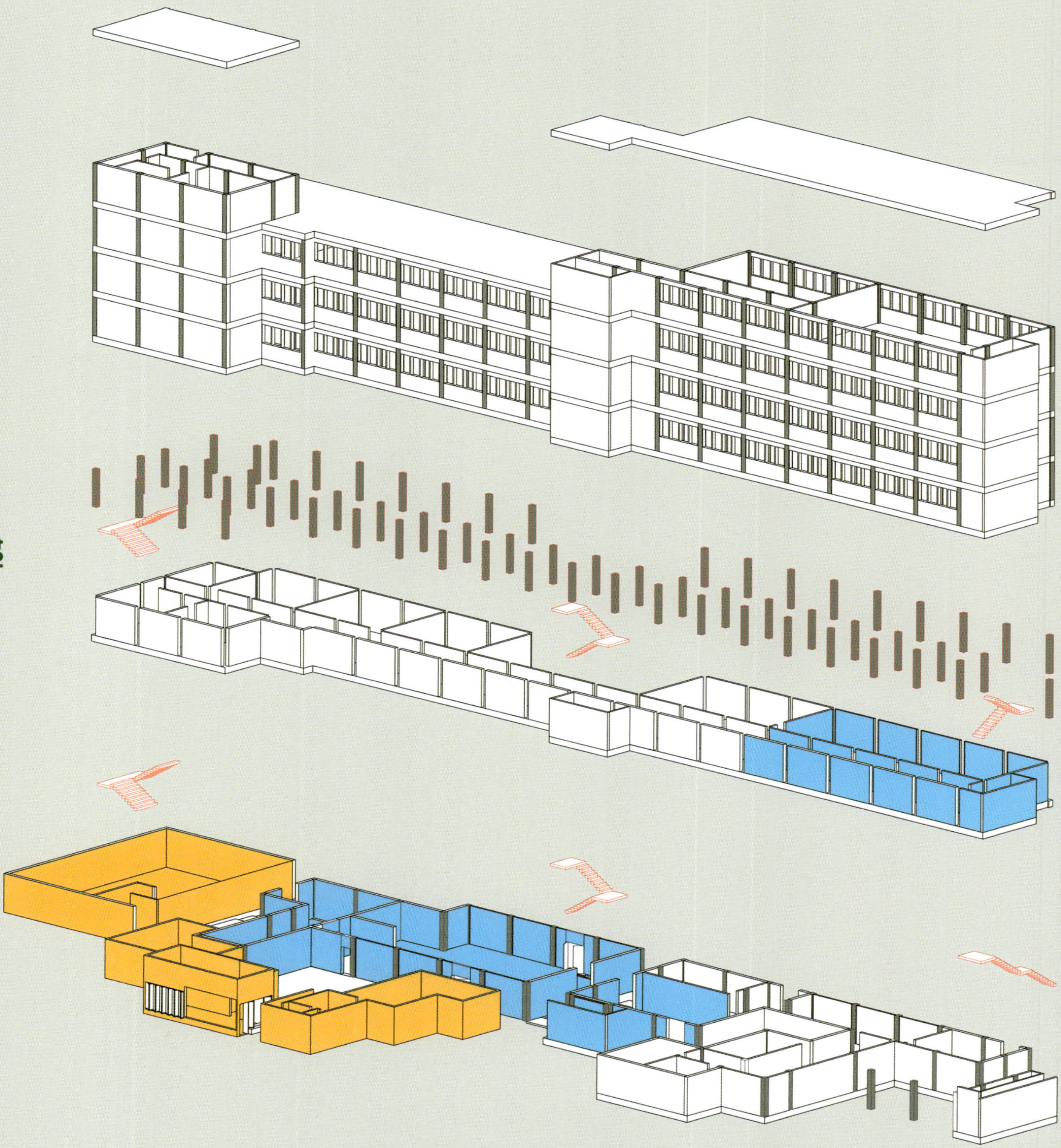

Newly Created Community Based Program (Residents)

Shared Program (Students+Residents)

School Program (Students)

Previous
Modification
1 Cafe
2 Laundry Room
3 Baths
4 Canteen
Cooking Studio
5 Printing Room
6 Administrative Office
7 Principal's Office
8 Staff Office
9 Sports Facilities Storage
1
3
5
10
20

서울이라는 도시는 역사와 현대를 아우르는 여러가지 레이어들이 중첩되어,
우리가 외국에서 흔히 보는 그리드형 질서정연한 필지와 가로와는 거리가 멀다.
같은 맥락으로, 송현동 부지는 남쪽으로 왕복 8차선 율곡로를 마주보고 있고 북촌과 경복궁 등의
역사적 의의가 있는 장소들로 둘러 쌓여 있다.
특히 송현동 중심의 덕성여자중학교를 기점으로 거주민과 외부인이 충돌하는
이질적 모습을 관찰할 수 있다. 해당 프로젝트는 침범 당한 거주민의 영역을 완화시켜주고자
기존 프로그램인 중학교와 공동체의 역할을 하는 동네 목욕탕을 하나의 마을로 확장시켜
근린의 목적을 강화하고자 했다.

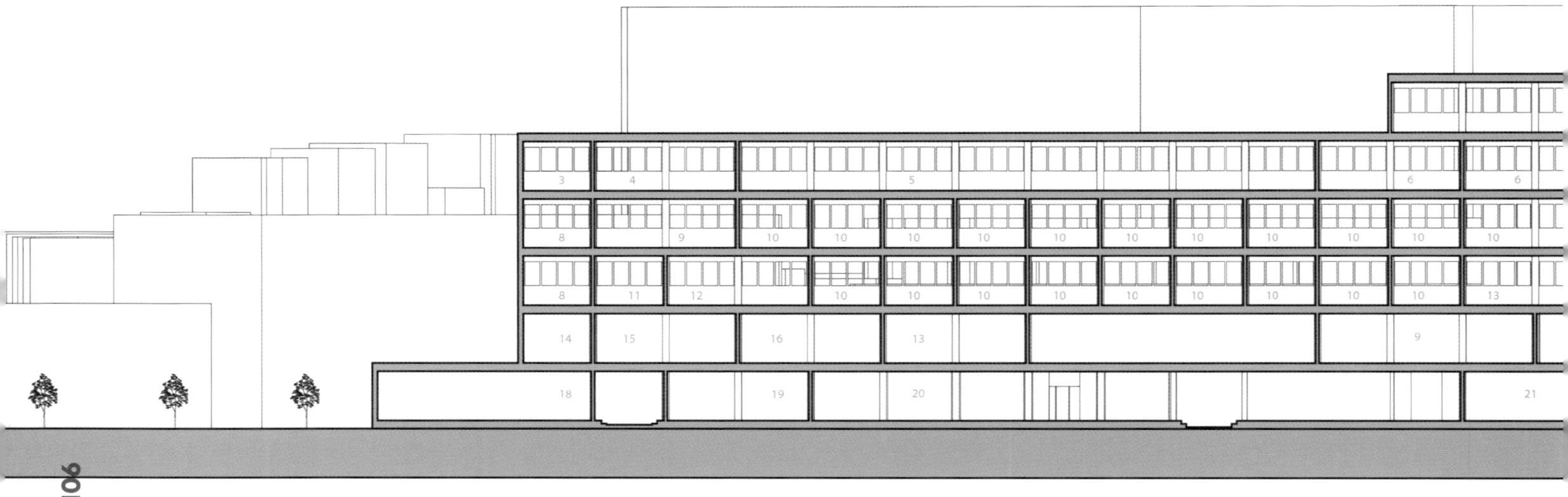

ivity Room 4 Music Room 7 After School Activity Room 10 Classroom 13 English Room 16 Broadcast Studio 19 Laundry Room 22 Staff's Office
ncing Studio 5 Auditorium 8 Student Lounge 11 Science Storage 14 Technical Room 17 Library 20 Canteen
sic Storage 6 Study Room 9 Computer Lab 12 Science Lab 15 Nurse's Office 18 Cafe 21 Printing Room

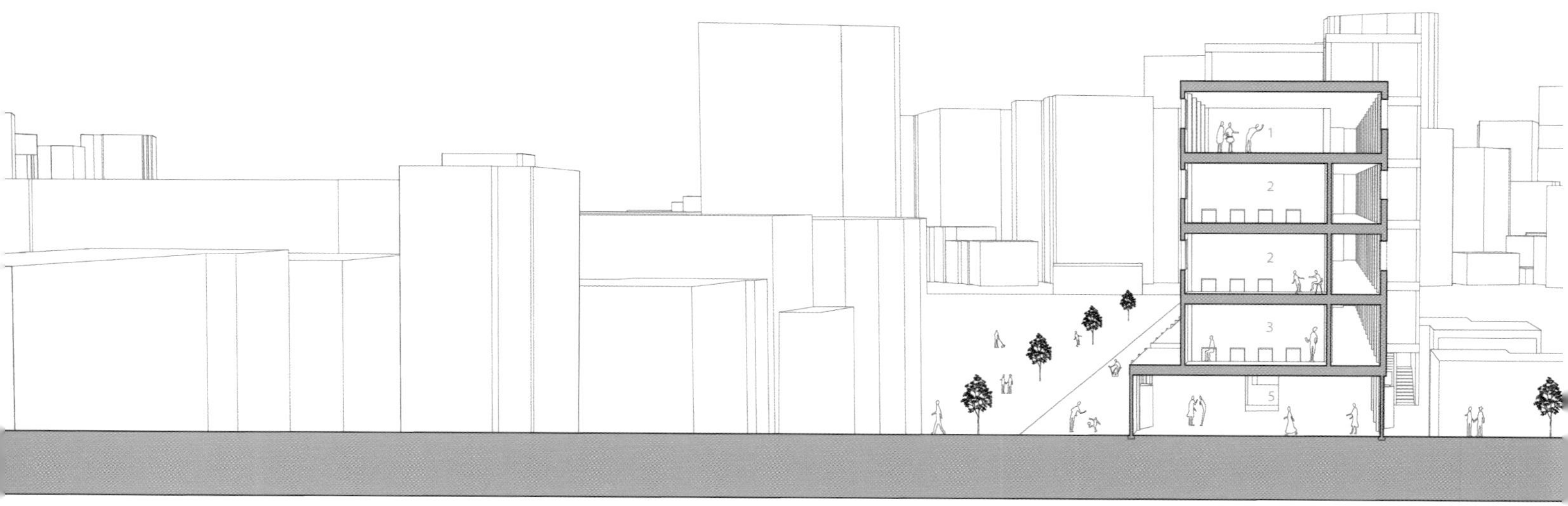

uditorium 2 Classroom 3 English Room 4 Canteen/Cooking Hall

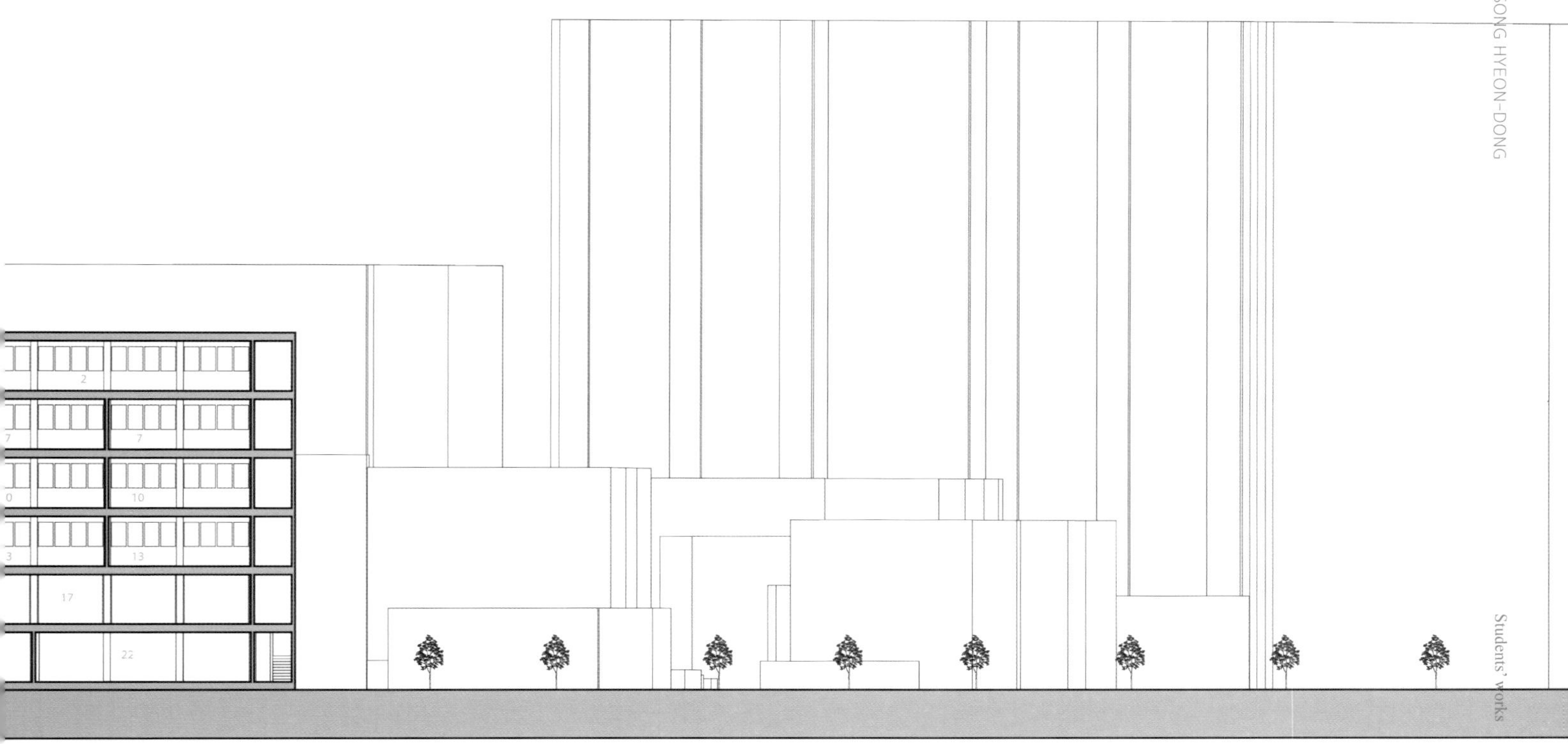

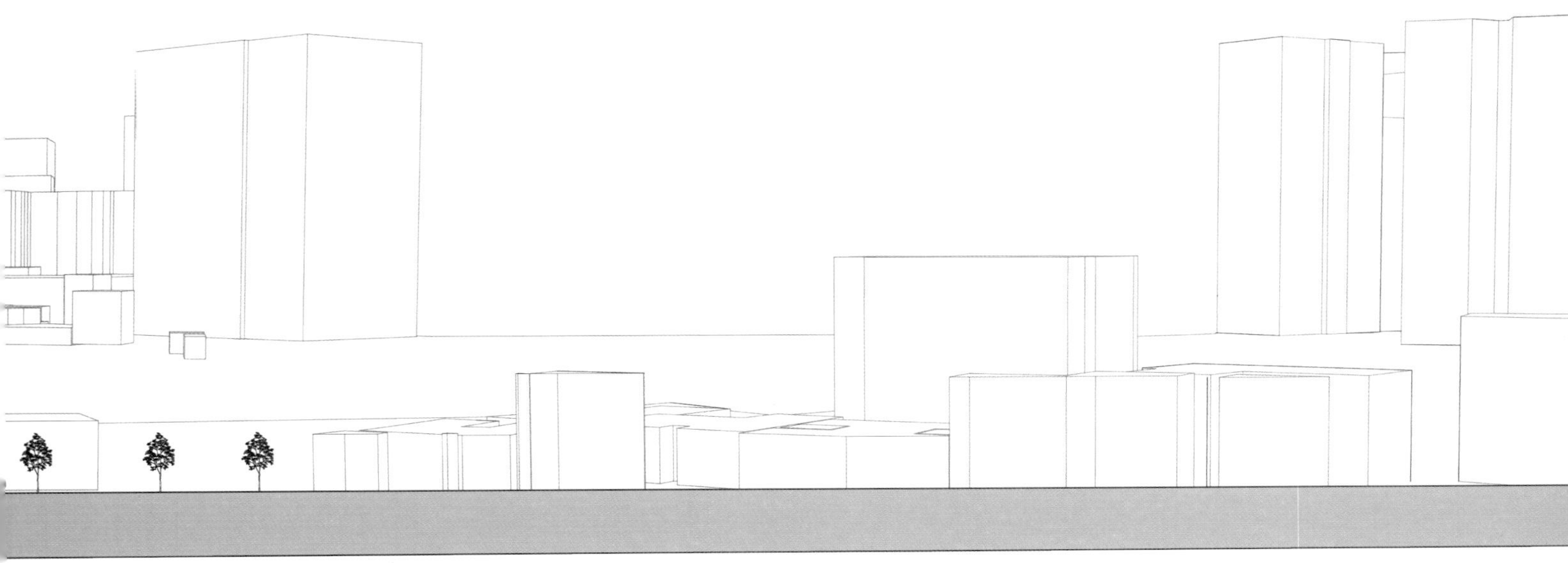

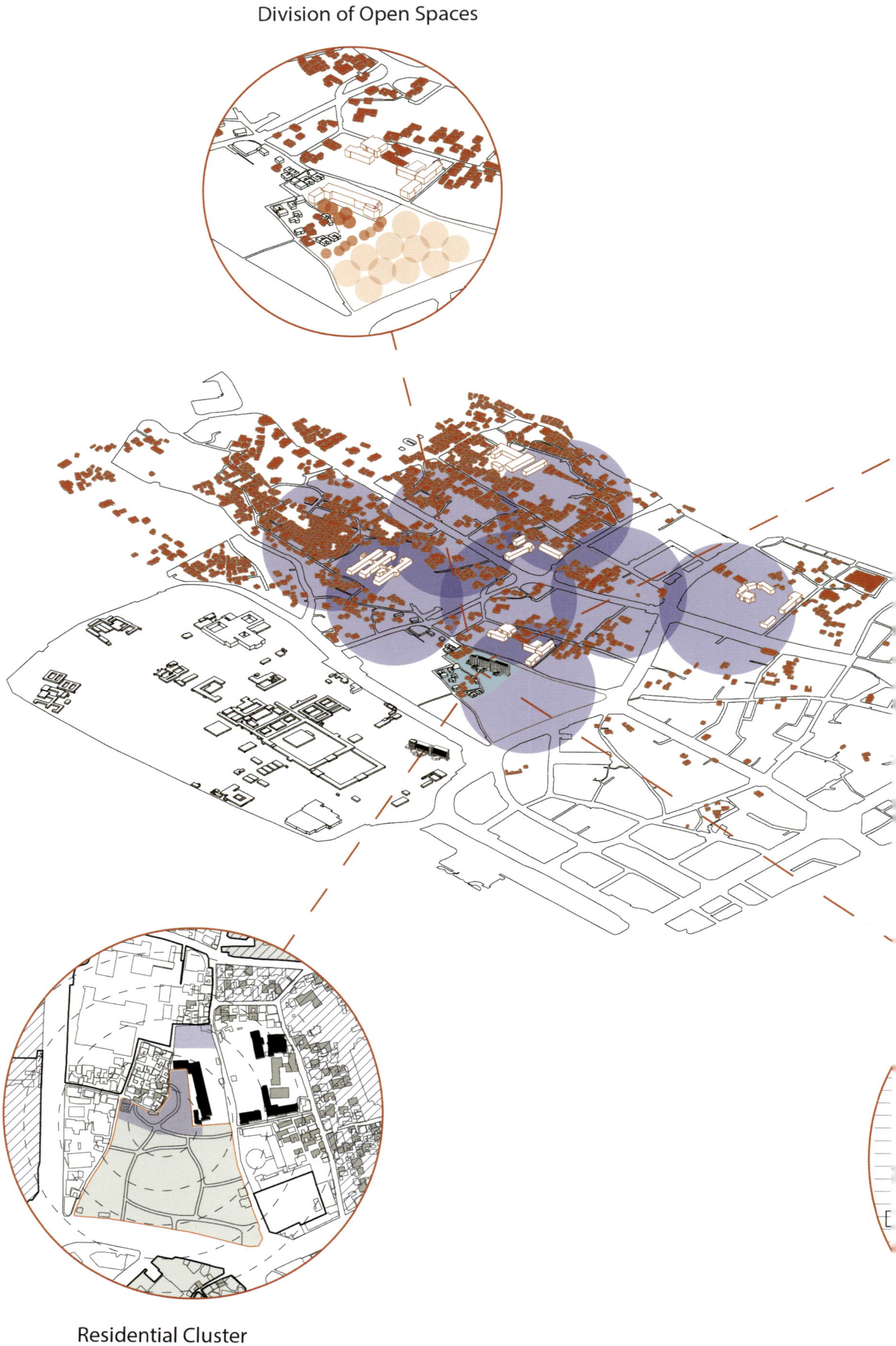
Division of Open Spaces
Residential Cluster

New Opportunities

Initiative:
Seoul, a city characterized by layered juxtapositions spanning historical eras to the contemporary, finds a pronounced manifestation of these intricate layers in Songhyeon dong. This locale boldly showcases collisions between diverse elements, such as the convergence of modern architectural edifices with traditional hanok buildings esteemed for their cultural significance. Previous research delved deeply into dissecting these juxtaposed layers, segmenting them into three facets reflective of various perspectives.

The focal point of collision investigated lay at the heart of Deok-seong Middle School, where residents and temporary visitors or tourists intermingled. However, prior to an in-depth exploration of this collision, there's a need to collectively redefine the term "collision." Conventionally carrying a negative connotation, it of-ten implies conflicting groups, potentially hinting at one group's dominance and an ensuing imbalance.

Interestingly, the site in question is enveloped by an array of cultural landmarks and attractions nurturing tourism and fostering gentrification in its vicinity. Consequently, the overarching objective of the project aimed to rectify this imbalance by expanding spaces dedicated to residents. This endeavor involved the delineation of purpose-driven open spaces and the integration of communal baths within the program, attempting to harmonize the flow within daily-use spaces.

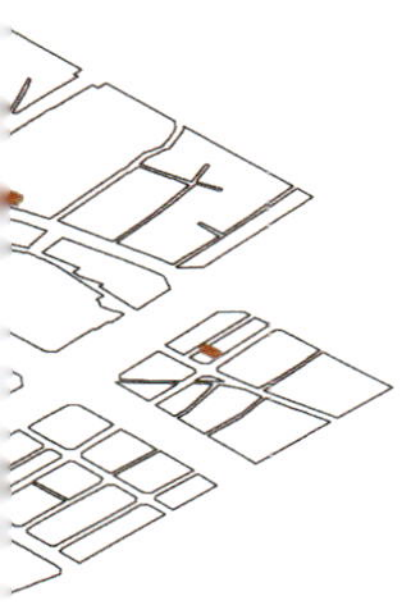

Program- Why Baths
Bathscape: Weaving Collective Experiences
Bathhouses, functioning as a third space, possess a multifaceted significance that transcends their utilitarian purpose. Beyond being mere facilities for bodily cleansing, they assume a pivotal role as communal hubs, where societal norms of social stratification and racial distinctions are subdued. Within these spaces, individuals from diverse backgrounds converge, enabling an environment where socioeconomic status and racial barriers lose their potency. This dissolution of barriers fosters an egalitarian ethos, allowing for the convergence of various social strata and ethnicities on neutral ground.

The bathhouse, in its essence, embodies a microcosm of society, where the act of communal bathing becomes a conduit for social interaction and discourse. Strangers, often separated by societal constructs, converge here, engaging in dialogues that transcend the limitations imposed by their external identities. Conversations unfold, opinions are shared, and connections form, facilitated by the shared vulnerability and intimacy inherent in the act of cleansing oneself. This shared vulnerability creates a unique platform for intermingling, enabling the convergence of disparate perspectives, experiences, and narratives.

Brief History
Public baths emerged due to the scarcity of private bathing facilities in densely populated areas, serving a primary function of communal hygiene. Despite being labeled as public, these spaces often enforced restrictions based on gender, religious affiliations, personal memberships, and other criteria. However, beyond their hygiene-oriented role, public baths evolved into significant social meeting points.

Scale of Local Baths

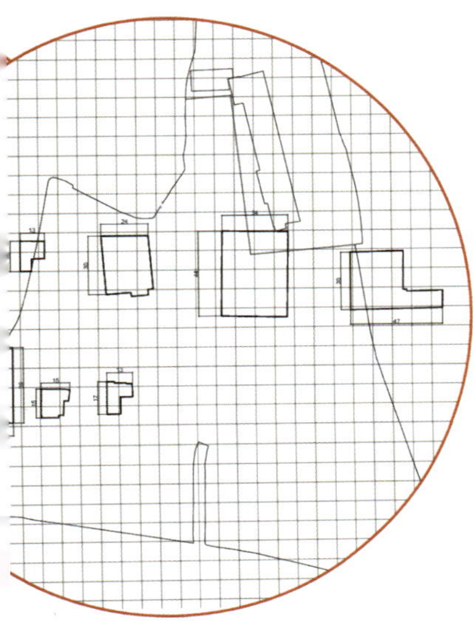

As societies witnessed an increase in households equipped with private bathrooms, the necessity for public baths declined, leading these spaces to predominantly serve recreational purposes. Notably, in Korean culture, public baths are renowned for their diverse amenities, extending beyond fundamental bathing facilities.
These encompass a wide spectrum of offerings, such as public saunas, hot tubs, showers, and massage tables for patrons seeking scrub massages, elevating the experience beyond mere hygiene to encompass relaxation and wellness elements.

Program-> Explanation
With the hopes to utilize the school as a basis for community experience, efforts were made to distinguish between altered and unchanged elements of the school.
Floors from third to sixth were kept as is, sustaining the school's educational purpose.
In contrast, floors on the ground level and the second level, colored in blue identify as shared spaces where both students and residents can interchangeably use.
Masses in orange identify as programs that are newly substituted that convey more of a resident-based function.

The Hidden History

The expansive site of Songhyeon-dong is now open to public as an urban green space following over 100 years of closure.

During the Japanese Colonia era from 1910 to 1945, the site was occupied by the Japanese and used as housing for Joseon Shiksan Bank, the capital power behind Japanese colonialism.

Since liberation to 1997, the site housed US diplomats. Samsung Life Insurance subsequently acquired the land in 1997 and sold to Korean Air in 2008.

Korean Air went through multiple iterations of project proposal for the site, including a cultural complex, and most notoriously, a 7-star and Hanok inspired hotel. However, none of them was ever realized, mainly due to Korean Air's financial constraints and governmental regulations. In the 20 years possessed by the two mega corporations, the site was left vacant, undeveloped, and overgrown. In 2021, a tripartite agreement amongst Korean Air, the Seoul government, and Korea Land and Housing Corporation took place, granting the ownership transfer to the Seoul government.

By the end of 2022, the government transformed the site into an open public park with green lawns. The park will be open until 2024, when a museum project will be underway hosting the late Samsung Chairman Lee Kun-Hee's personal art collection.

The site is scheduled to reopen to the public upon the museum's anticipated completion in 2027, continue to serve as an urban green infrastructure and civic space in the city.

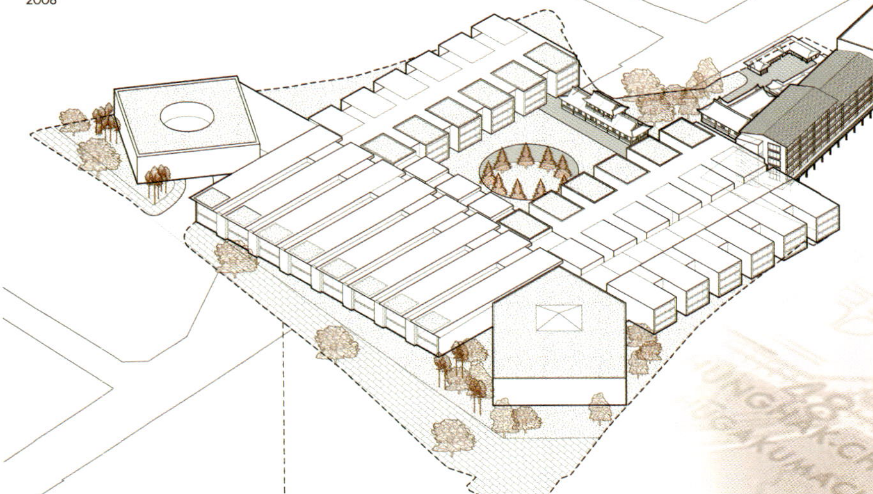

Korean Air purchased the Songhyeon-dong site from Samsung Life Insurance in 2009, with the intention of constructing the country's first 7-star hotel inspired by the form of the traditional Hanok. With the site's proximity to cultural and historic landmarks, the airline pushed for the hotel's value in bolstering tourism and cultural engagements. The airline commissioned Mario Botta, a renowned Swiss architect who designed a part of the Leeum museum, to come up with the concept design as illustrated. Nevertheless, the project proposal failed to proceed due to the restrictions of the School Health Law in Seoul barring the establishment of hotels around schools. There are three schools surrounding the site - Duksung Girl's Middle School, Duksung Girl's High School, and Pungmoon Girl's High School. The airline went through rounds of lobbying, attempting to leverage the government towards deregulation and promised investments in archaeological research and public engagement. When the hotel proposal fell through, the airline proposed a cultural center featuring recreational facilities and retail stores.

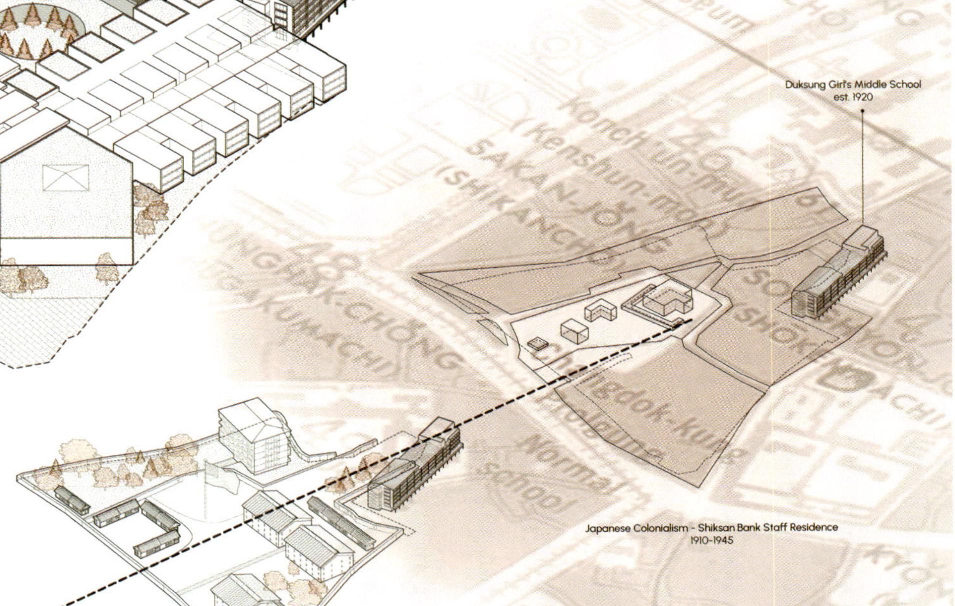

The Hidden History

The expansive site of Songhyeon-dong is now open to public as an urban green space following over 100 years of closure. During the Japanese Colonia era from 1910 to 1945, the site was occupied by the Japanese and used as housing for Joseon Shiksan Bank, the capital power behind Japanese colonialism. Since liberation to 1997, the site housed US diplomats. Samsung Life Insurance subsequently acquired the land in 1997 and sold to Korean Air in 2008. Korean Air went through multiple iterations of project proposal for the site, including a cultural complex, and most notoriously, a 7-star and Hanok inspired hotel. However, none of them was ever realized, mainly due to Korean Air's financial constraints and governmental regulations. In the 20 years possessed by the two mega corporations, the site was left vacant, undeveloped, and overgrown. In 2021, a tripartite agreement amongst Korean Air, the Seoul government, and Korea Land and Housing Corporation took place, granting the ownership transfer to the Seoul government. By the end of 2022, the government transformed the site into an open public park with green lawns. The park will be open until 2024, when a museum project will be underway hosting the late Samsung Chairman Lee Kun-Hee's personal art collection. The site is scheduled to reopen to the public upon the museum's anticipated completion in 2027, continue to serve as an urban green infrastructure and civic space in the city.

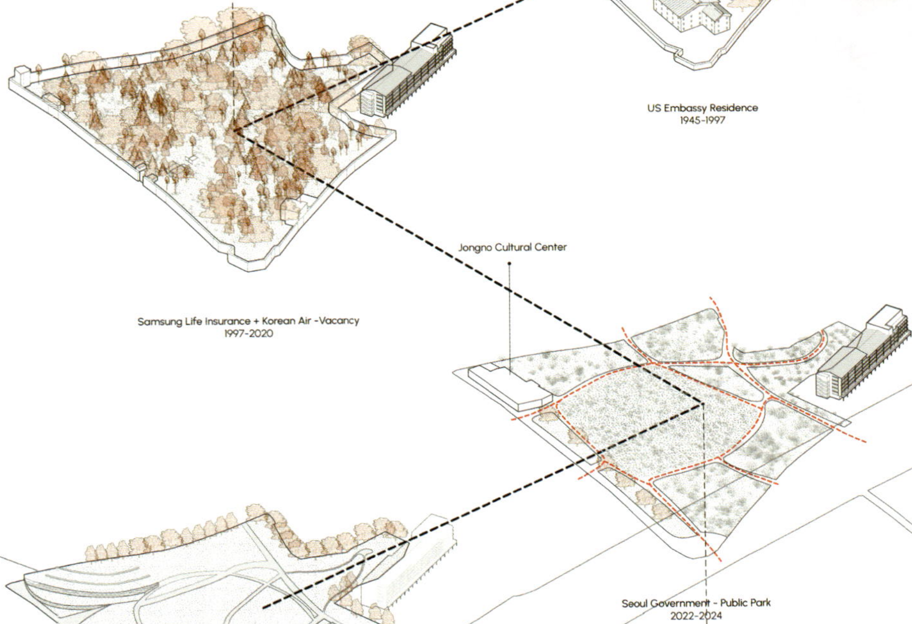

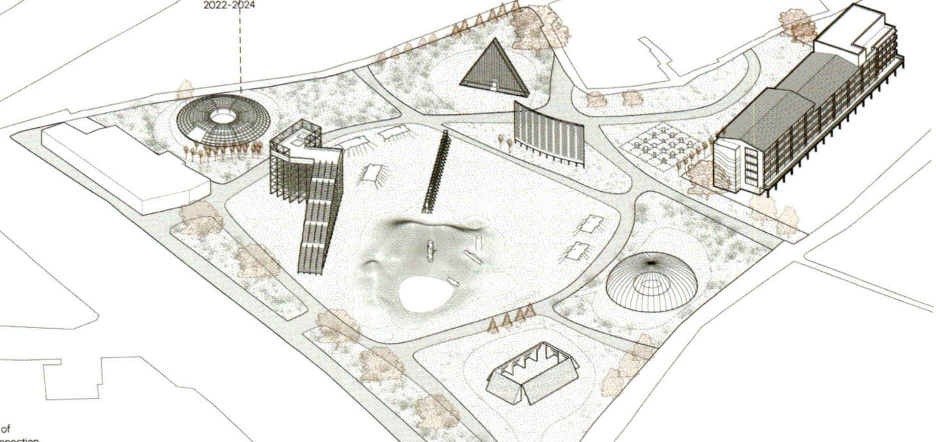

The 2023 Seoul Biennale of Architecture and Urbanism, titled "Land Architecture, Land Urbanism: City of Mountain Ranges, Waterways, and Wind Breezes - Drawing of Seoul's Next 100 Years," features the connection between the urban and the natural environment. With land and ecology at its core, the Biennale invites visitors to various pavilions in its main venue at Songhyeon-Dong. The temporary pavilions were built mainly using reusable and sustainable materials that can be easily assembled and dismantled. These projects explore the theme of an "Outdoor Room," inviting visitors to engage with built elements in open air and contemplating on the relationships between humans and nature.

BACK ALLEYS OF SEOUL

Traces of Pimagtol Alley are highlighted by notable commercial landmarks in Jongno, identified by locals who are nostalgic of old Seoul's character. Bordered by small eateries and craftstores, Pimagtol remained a sanctuary for the working class, who historically since the Joseon Era, used to cut through this parallel running 12 foot alley to avoid the busy wider main highways designated for royal officials.

The tension between demolition and the attempt on preservation is highlighted in the multi-use building 'Centropolis' and National Palace Museum of Seoul. Can the longing for these back alleys that now lie in the shadows of urban renewal be used to understand the need for cross-roads that facilate interaction and preserve heritage? Is there further need to displace this heritage, and house it in enclosed institutions scattered around the site?

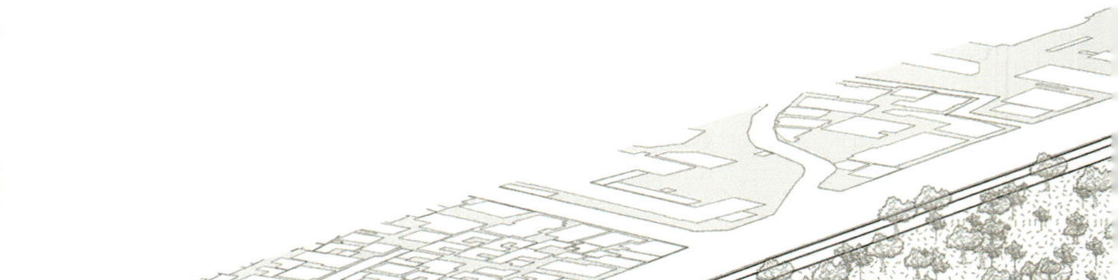

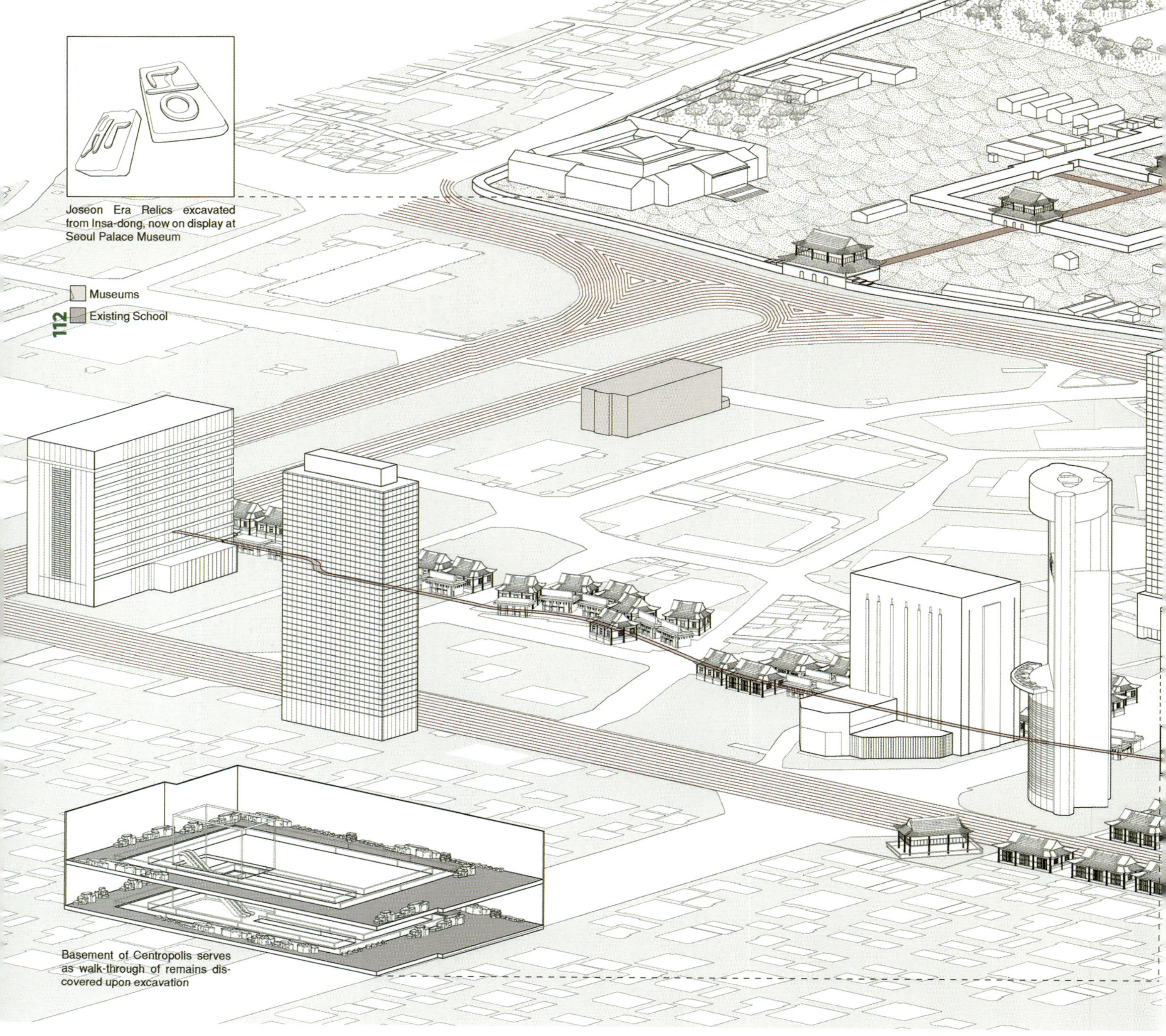

Joseon Era Relics excavated from Insa-dong, now on display at Seoul Palace Museum

Basement of Centropolis serves as walk-through of remains discovered upon excavation

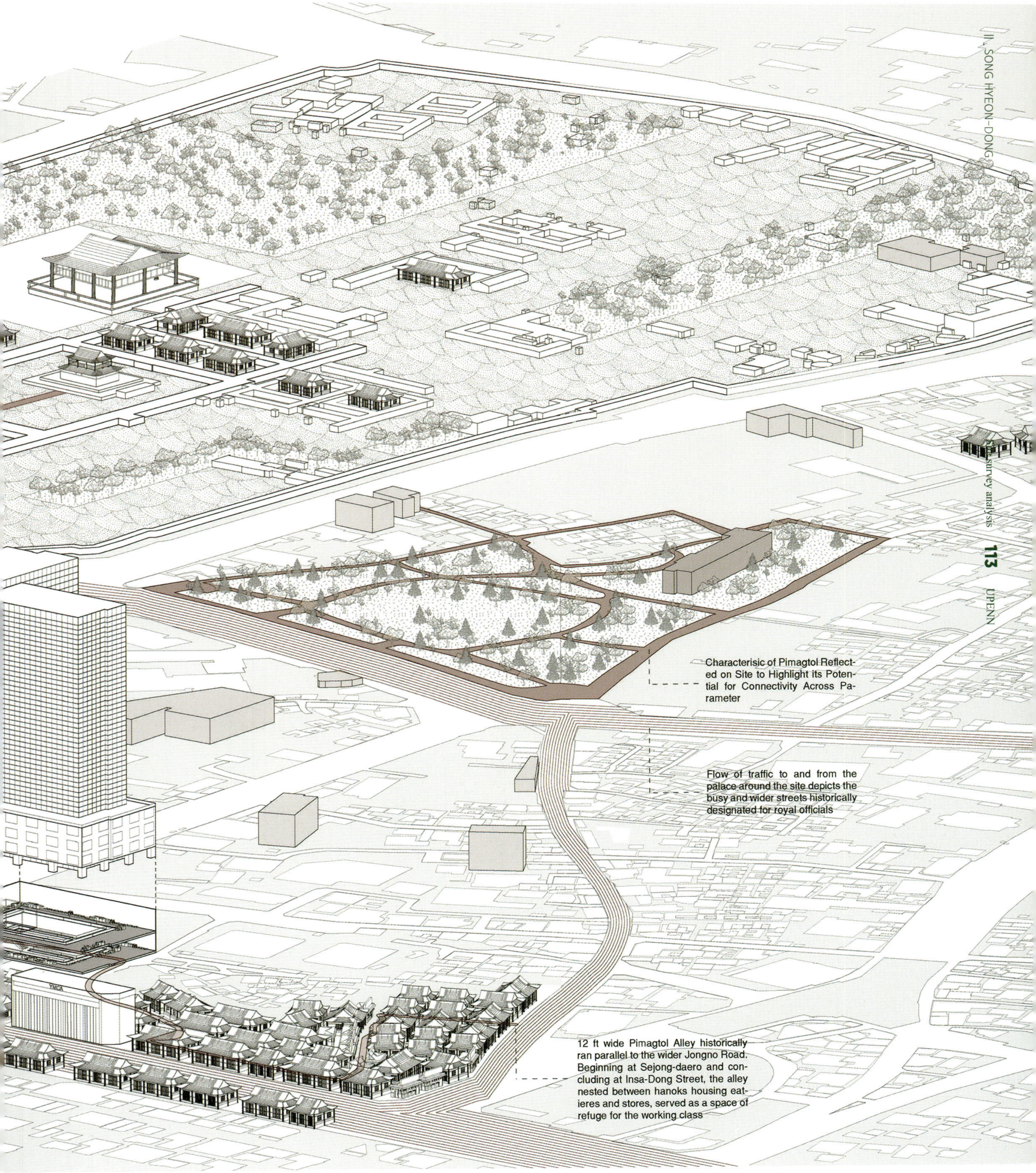
YMCA
Characterisic of Pimagtol Reflected on Site to Highlight its Potential for Connectivity Across Parameter
Flow of traffic to and from the palace around the site depicts the busy and wider streets historically designated for royal officials
12 ft wide Pimagtol Alley historically ran parallel to the wider Jongno Road. Beginning at Sejong-daero and concluding at Insa-Dong Street, the alley nested between hanoks housing eatieres and stores, served as a space of refuge for the working class

The urban pattern throughout history

Songhyeon-dong

The site in Songhyeon-dong has been used as a residence of Siksan Bank in Japanese colonial era.

The site in Songhyeon-dong has been used as A U.S. military residence, and a residence at the U.S. Embassy.

It was returned to the Korean government in 1997 and returned, but it has since been left in ruins without any use.

Now

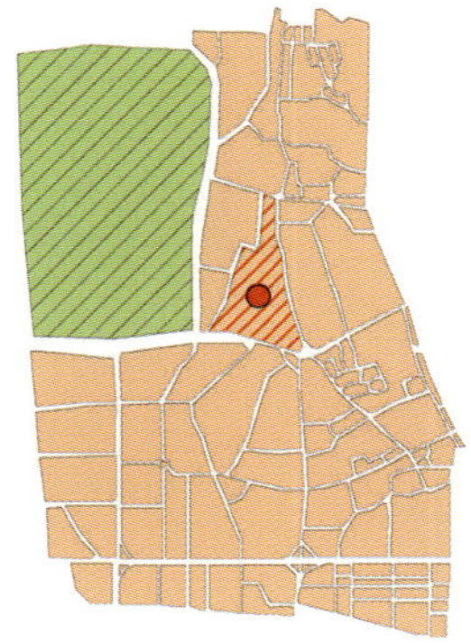

The 1914 Japanese Map of Hanseongbu

+

Th

Korea after liberation from Japanese rule in 1945 marked a significant turning point in its history. Relations between North and South Korea have been complex and at times hostile since the division. There have been periods of engagement and reconciliation, as well as periods of tension and conflict. Efforts to improve inter-Korean relations continue to be a prominent feature of both nations' foreign policies. The post-liberation period in Korea is marked by both remarkable achievements and ongoing challenges, including the division of the peninsula, efforts toward reunification, and the pursuit of peace in the region.

After Liberation

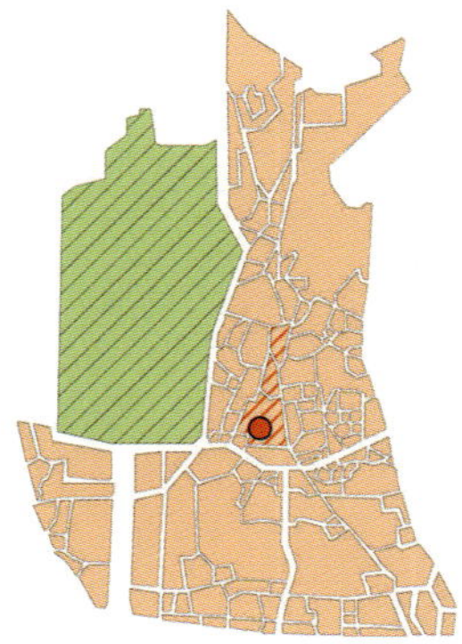

Korea under Japanese rule refers to the period of time from 1910 to 1945 when the Korean Peninsula was a colony of the Empire of Japan. During this period, Korea underwent significant political, social, economic, and cultural changes, with Japan exercising colonial control over the region. The period of Japanese colonial rule in Korea is remembered as a difficult and challenging time in Korean history. It left a complex legacy that continues to shape modern Korean identity, and it is often a subject of historical reflection and discussion in both South and North Korea.

Under Japanese Rule

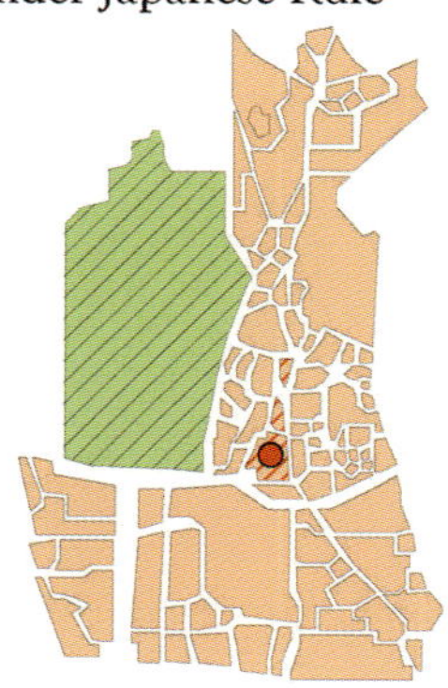

Shinsen chosenkoku zenzu (Newly Edited Whole Map of Joseon Dynasty) by Tanaka

The founding of the Joseon Dynasty had significant implications for Korean history and culture. During its rule, the dynasty implemented a number of important reforms, including the adoption of Confucianism as the state ideology and the creation of a centralized bureaucracy. These reforms had a lasting impact on Korean society and governance, shaping the country's political and cultural landscape for centuries.

Joseon Dynasty

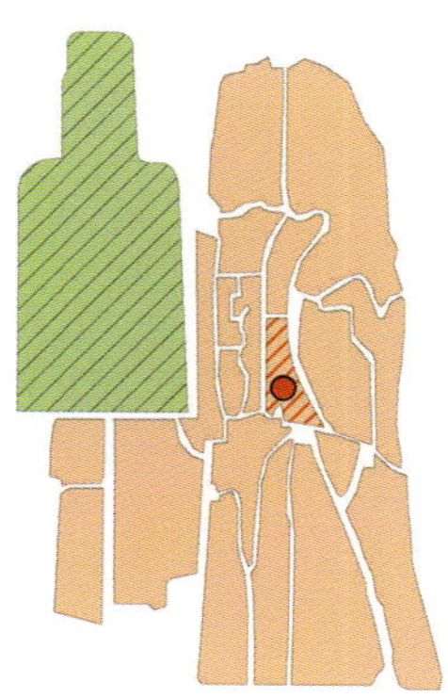

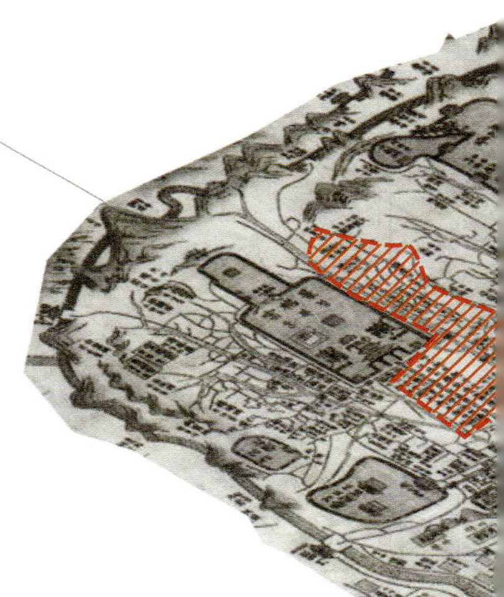

Standardized Street Map Based on the 1914 Map

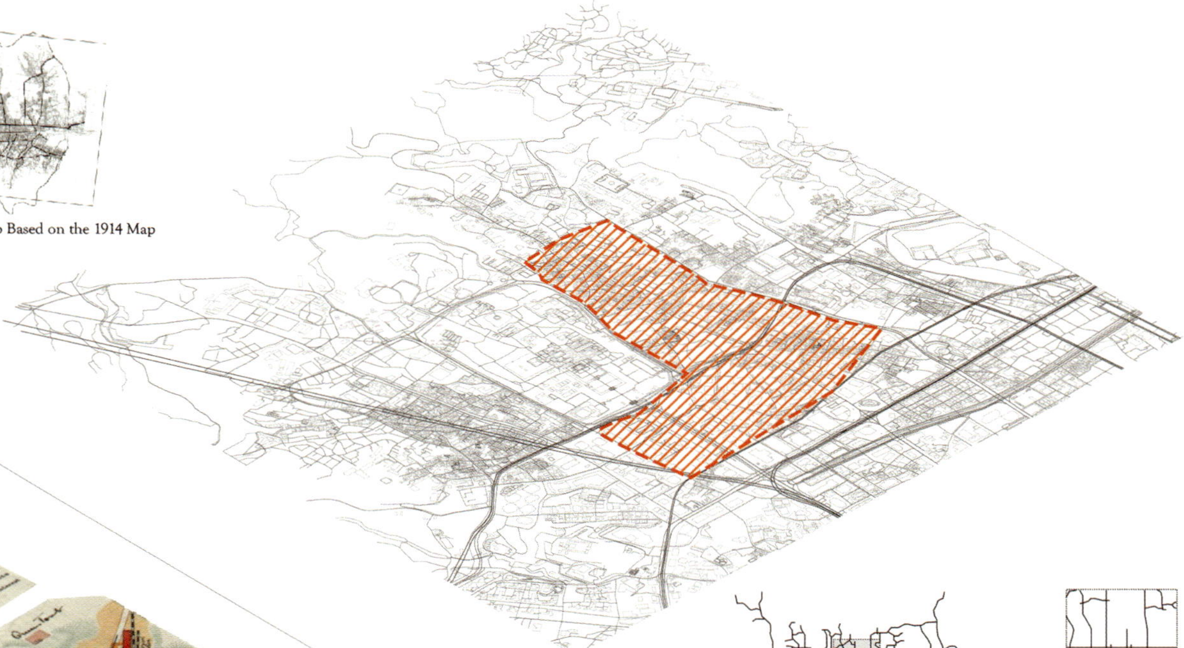

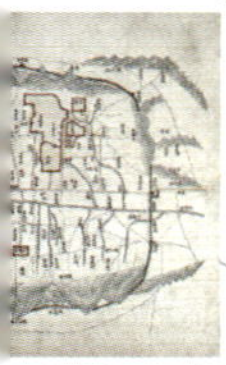

ojido Map of 1861

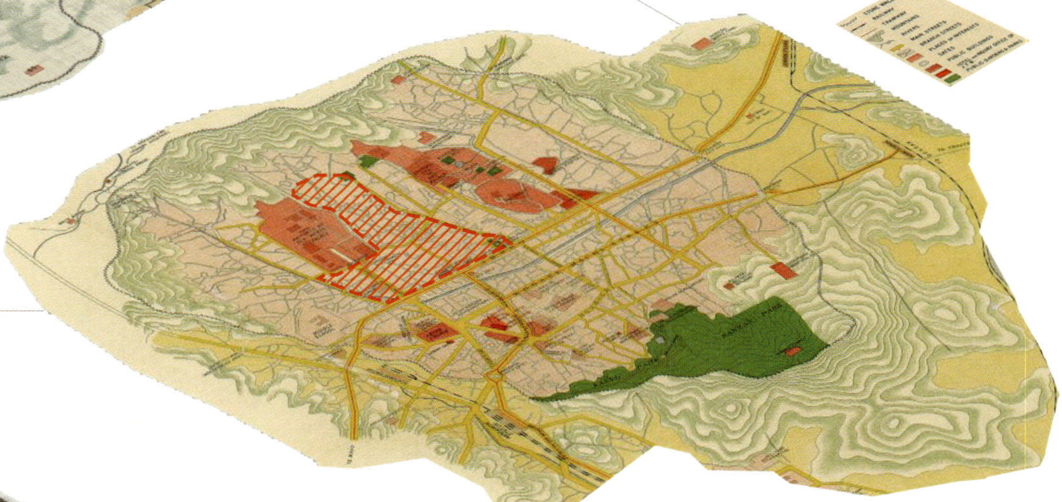

The Dosung Samkunmun Bungoejido Map of 1751 with Curved Lines

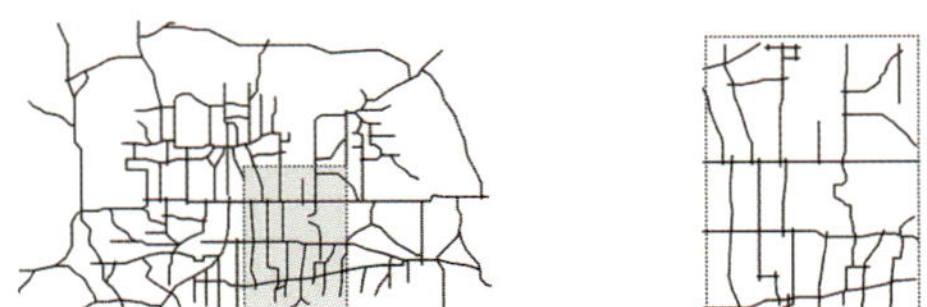

An Axial Map of the Daedong Yojido Map of 1861 with Straight Lines

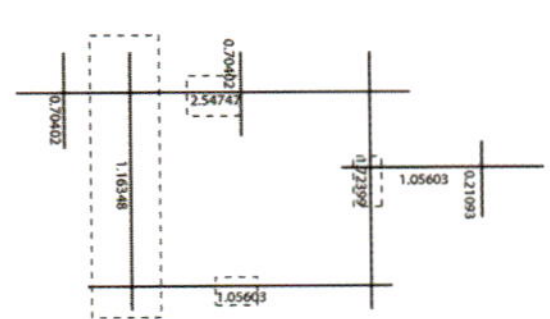

Integration Values for a Map with Straight Lines

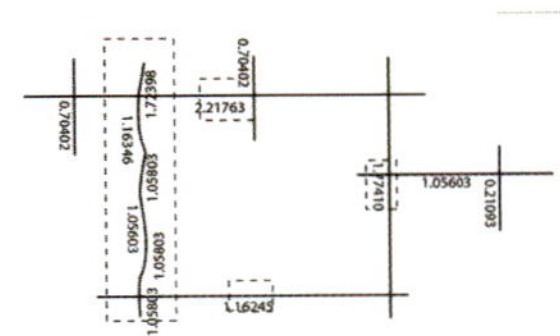

Integration Values for a Map with Curved Lines

POPULATION//LAND//DEMOGRAPHICS

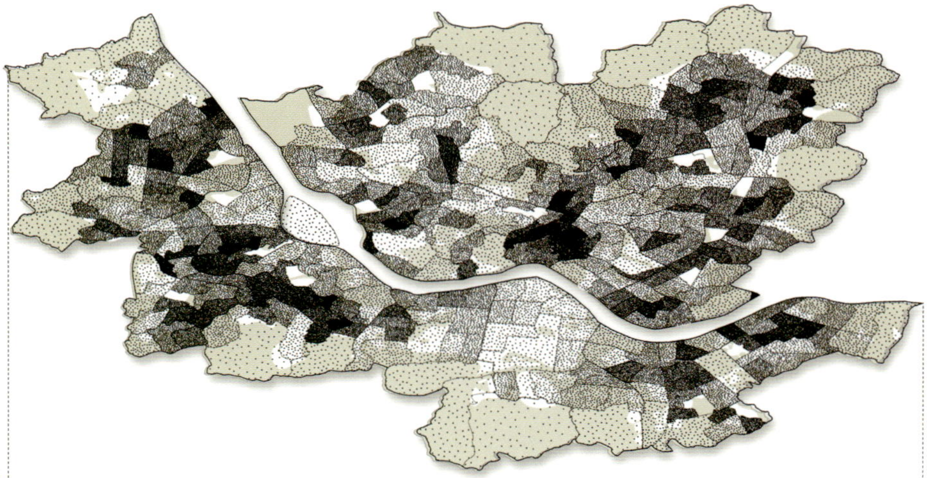

2040: POPULATIONDECREASE, MEDIAN AGE 50

2010: PEAK POPULATION DENSITY, MEDIAN AGE = 36.6

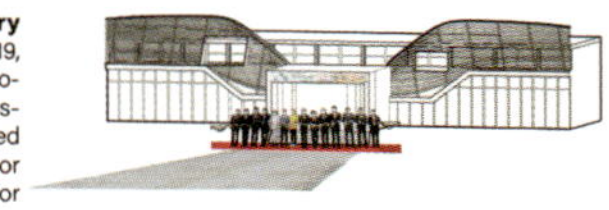

Soungsan Elementary School: Shut down in 2019, this school, in the South Gyeongsang Province, is now transformed into a jungle themed children's library with outdoor camping facilities. The interior space is also used for lectures and performances.

Nogok Primary School: In spring of 2016, the school closed down with a graduating class of 12 students. The school, once a symbol of progress within an idyllic town among the hills now bears the scars of rapid industrialization and dwindling populations in the country side. The county has seen no new births in the past several years.

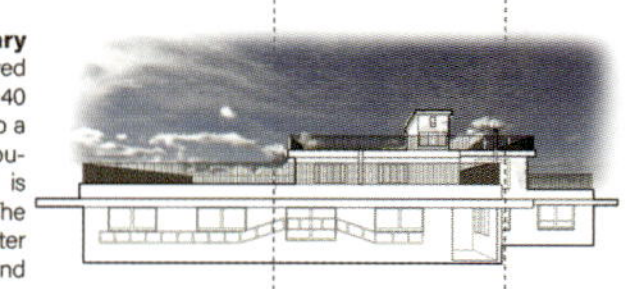

Hansan Elementary School: This school served the community from 1940 until its close in 2012 due to a drastic loss of student population. Now, the school is reborn as a "cat school." The building will provide shelter and care for stray cats and generate regional income through the sale of pet goods and foods.

2010: PEAK POPULATION DENSITY, MEDIAN AGE = 36.6

1975: FULL INDUSTRIALIZATION REALIZED, MEDIAN AGE = 18.8

1915: AGRARIAN CITY, MEDIAN AGE ≈ 17

Urban Hardscape

Forest/Agriculture

Population Density

Nokodo Elementary School: Closed in 2006, the three remaining children on Nokodo Island are tutored by a teacher who arrives by boat. In order to give hope to the community, the father of the children hopes to give his children a middle school education on the island. Currently the school and its grounds sitsabandoned and are falling into dissaray.

Chung-il Girl's High School: Closed in 2006, this former school is reportedly haunted. The building is a 5 story building that at one point housed over 3,000 students. The school opened in 1979 during a boom from the textile industry. The school meant to provide young people of the area a promising future in the industry, however, the boom did not las. By the closing of the school only 21 students remained. The school and the grounds are left deserted.

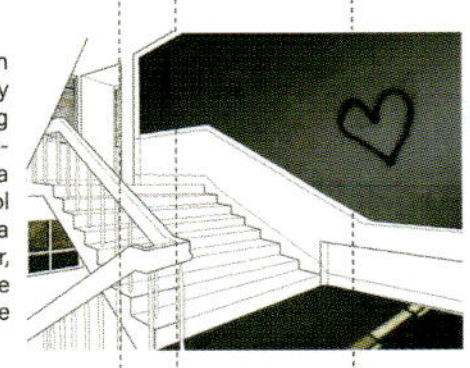

Population Growth: above 1

Population Growth: -1, 1

Population Growth: below -1

Reprogrammable City
Songhyeon-dong Reimagination

Aimen Arain & Reggie Zhao

Our initial research focused on Seoul's dynamic culture driven by the idea of a "Lack of Destruction." Layers of history stemming from the Joseon Era rule, paired with this societal gravitation to "progress" through a quick shift towards mid century modernity after the war, resulted in the drastic growth of industrial commercial projects.
The idea of neither ruin nor revival was prevalent as Seoul transitioned through mid century changes, and the tension between 'growth' within the built environment in the midst of urban historical decay is prevalent in present day Seoul.

Instances of societal refuge are prevalent within the culture through many forms. Alleys throughout Seoul serve as intimate parameters for connection, an escape from the fast-paced commercial take over. A particular alley we chose to highlight is the historic 'Pimagtol' Alley, that runs parallel to the main Jongno Road.
The 12-ft alley historically served as a separate gathering space for the working class that served royal officials. Hanoks and storefronts line the alley, and it became a space that cultivated the exchange of community engagement and culture.
Today traces of Pimagtol lie, as most of it was demolished to make room for notable commercial buildings. Some of these buildings such as Centropolis preserved the historic sites discovered through excavation and serve as an example of the tension between decay and 'growth'.

The concept of refuge is also prevalent in the Bang "room" culture.
Jjimjilbangs are bathhouses that offer additional communal spaces for community engagement, while also offering facilities such as arcades, movie theaters, gym or karaoke.
The bang room culture is popular in Seoul as it offers designated space for people to participate in activities in a contained space away from where they may reside.

The project proposal on the Songhyeon-dong site aimed to maintain its existing qualities of being an open air green space surrounded by significantly different programs on either side of the site. The target demographic is the pre-college year youth and the proposal includes temporary housing for the displaced youth community.
The existing school on site is repurposed as rooms for education and learning that connect to clusters that house 'bangs' ranging from art to saunas, sports, community kitchens, cafes, gardening, dance and performance.
The idea of connection throughout the site paired with two clusters that offer a variety of 'outdoor' bangs, was inspired by Seoul's cultural typology while adhering to the site's existing potential to be a less dense space for gathering and engagement.

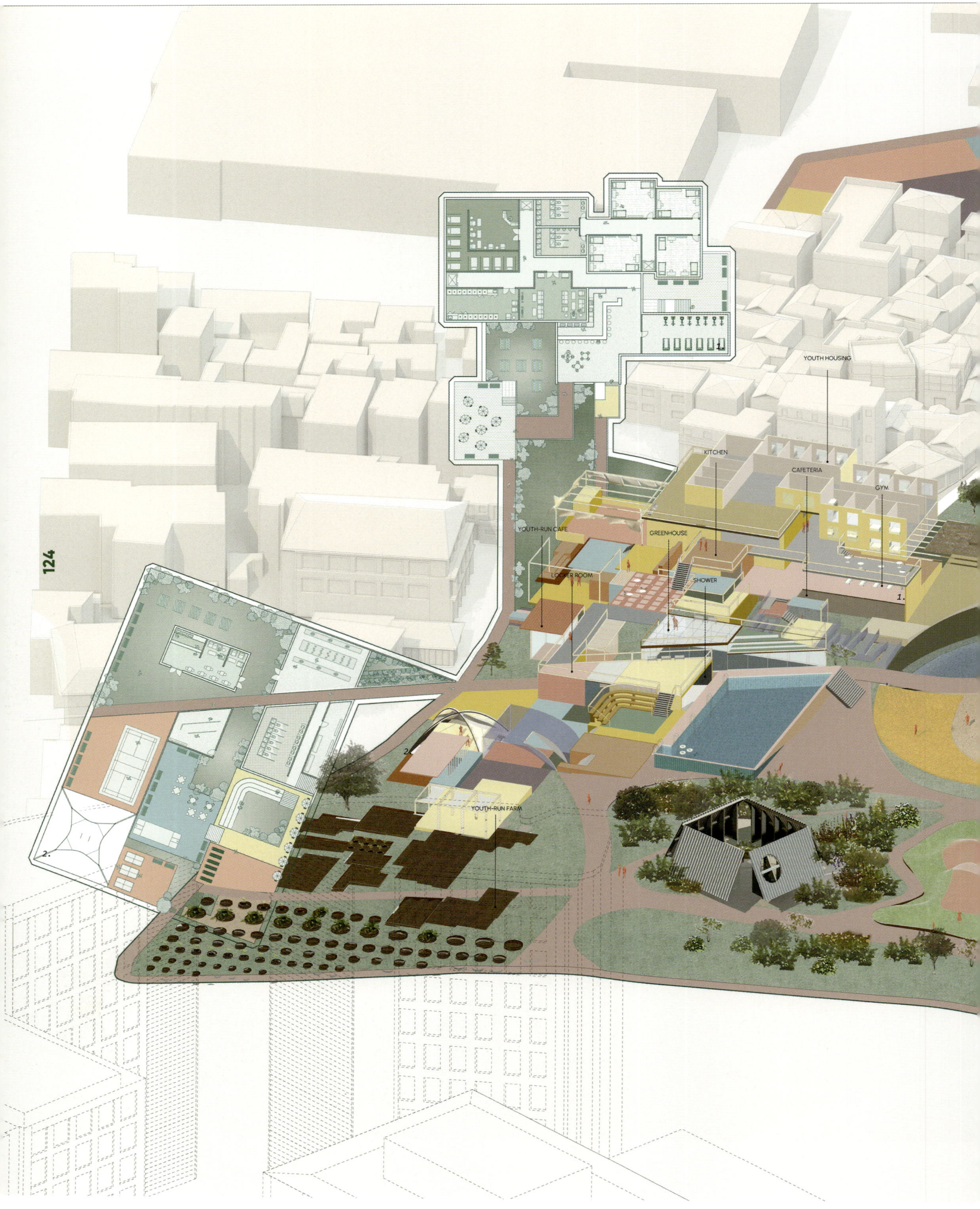

1.
YOUTH HOUSING
KITCHEN
CAFETERIA
GYM
YOUTH-RUN CAFE
GREENHOUSE
LOCKER ROOM
SHOWER
1.
2.
YOUTH-RUN FARM
2.

GOSHIWON
(EXAM PREP HOUSING)
EDUCATION CENTER
EDUCATION CENTER
PC BANG
YOUTH HOSTEL
LOUNGE
COMMUNITY KITCHEN
YOUTH HOSTEL
ROOMS
NORAEBANG
(KARAOKE)
GALLERY
CONVENIENCE STORE
JIMJILBANG
(BATHHOUSE)
3.
3.

Project Blurb

Looking through the lens of degrowth and intergenerational care, our project analyses the urban history of the Songhyeon-dong site and proposes a way to preserve its sense of relief within the dense urban fabric while outlining various programs which allow for the formal and informal engagement of people of all ages.

As the city's population continues to shrink and age, and more buildings become vacant, the proposed interventions outlined here may serve as a model to address newly available vacancies in service of the needs of the public, focusing on the growing elderly and shrinking youth populations.

RESOURCE CENTER
MAKER +
GALLERY SPACE
MENTAL HEALTH
+ MEDITATION CENTER
SUBWAY
ENTRANCE/EXIT

ON
RENNOVATED SCHOOL BUILDING
POETRY + WRITING CENTER

RESOURCE CENTER
MAKER +
GALLERY SPACE
MENTAL HEALTH
+ MEDITATION CENTER
SUBWAY
ENTRANCE/EXIT

RENNOVATED
SCHOOL BUILDING
POETRY +
WRITING CENTER

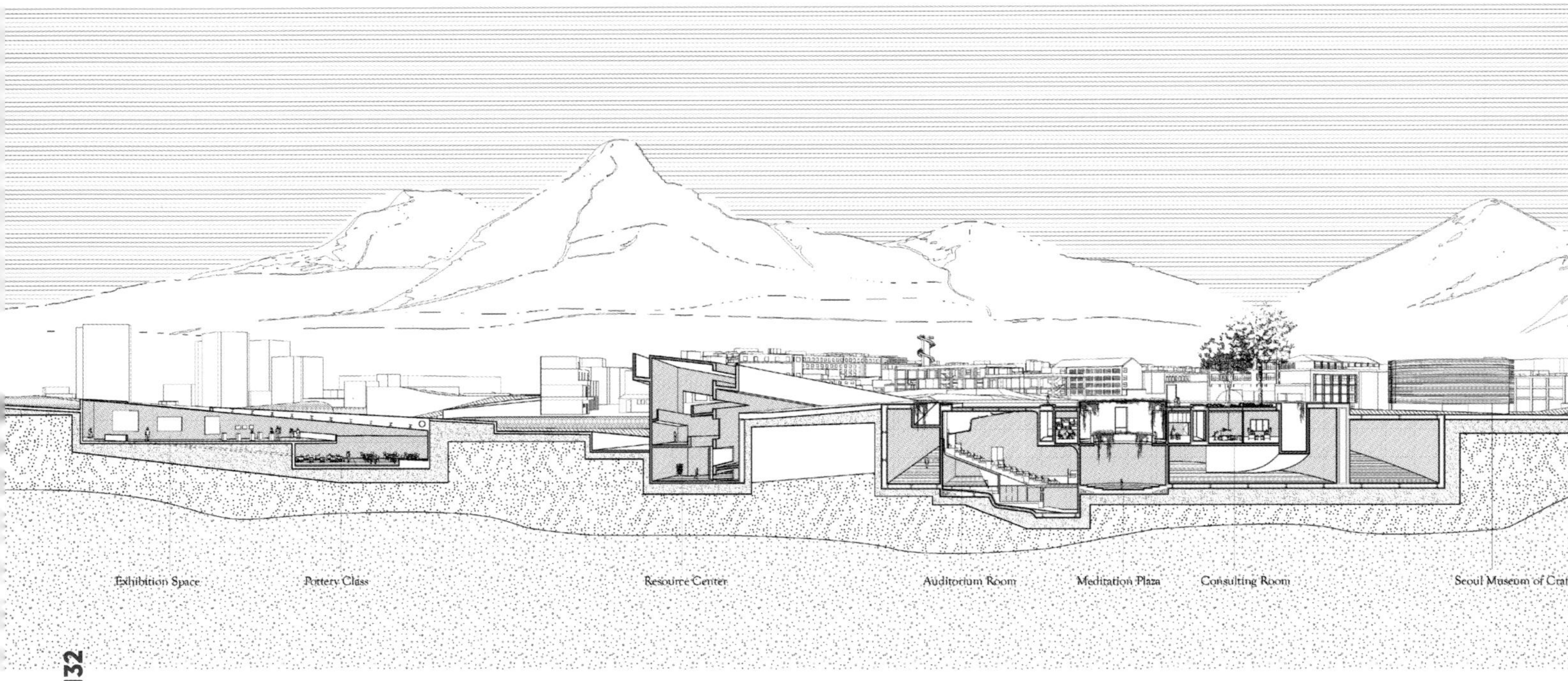
Exhibition Space
Pottery Class
Resource Center
Auditorium Room
Meditation Plaza
Consulting Room
Seoul Museum of Craf

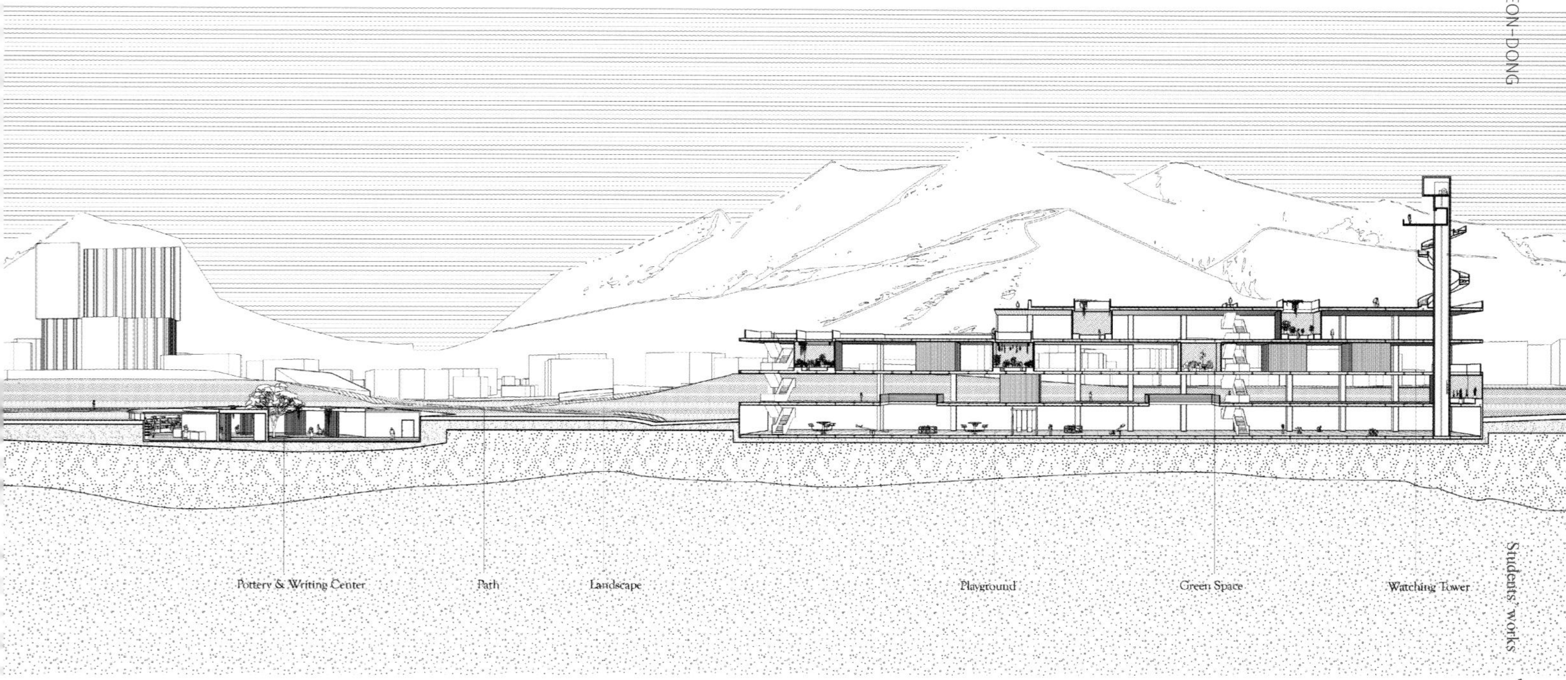

UPENN _ Amy Greenberger + Pamela Pan

THE REPROGRAMMABLE CITY, SEOUL

TERRA INCOGNITA

FAR EAST BASE

TERRAINCOGNITA

III

FAR EAST BAS

60년간 미군 기지사용 되었던 미 극동 공병단 부지는 '지도에도 없는 땅'으로 주변 도시 맥락과 단절되어 있다.

The site has been used as a military base of U.S. Army Far East District for 60 years and is disconnected from the urban context around it. It is a Terra Incognita.

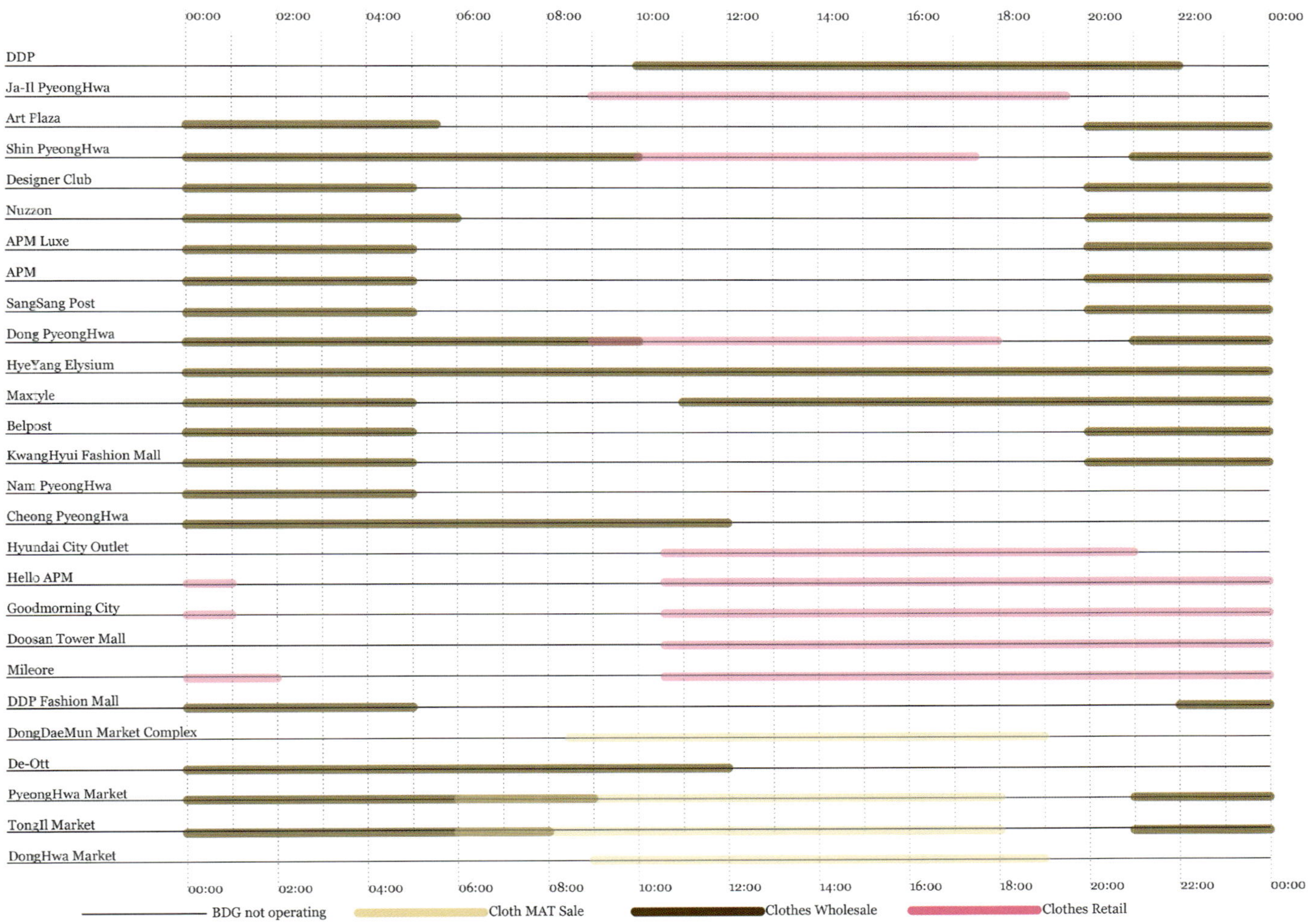

사이트 주변은 동대문 패션 디자인 특구이다. 24시간 작동하는 특구를 패션 상가의 운영시간을 통해 확인해볼 수 있었다. 낮 시간대에는 소매상점이, 밤 시간대에는 도매상점이 주로 운영되며 서로 다른 시간대에 서로 다른 이용자를 끌어들인다.

The area around the site is the Dongdaemun Fashion Design District. Through the operating hours of fashion shops, we could observe the district operating 24 hours. Retails open at daytime, Wholesales open at nighttime resulting in different occupants in the area at different time of a day.

훈련도감, 1392 – 1919
Military Training Camp, 1392 – 1919

도성대지도, 18세기 후반
Hanyang city map, late 18th century

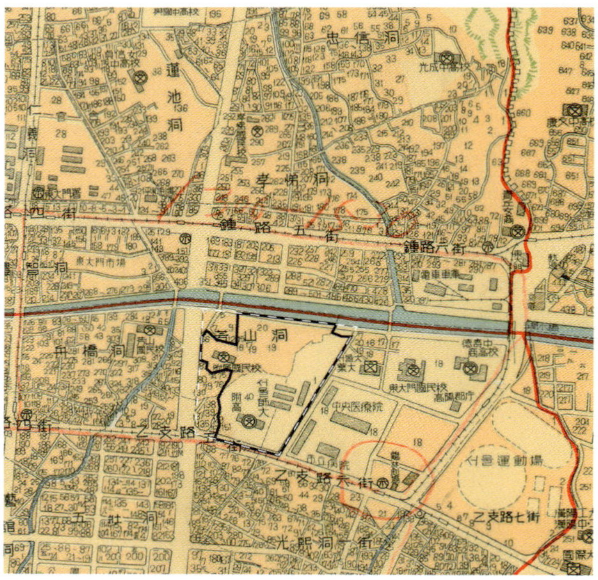

경섬사범학교, 1922 – 1946
Gyeongseong School of Education, 1922 – 1946

지번입서울특별시가지도, 1958
Seoul Metropolitan Government map, 1958

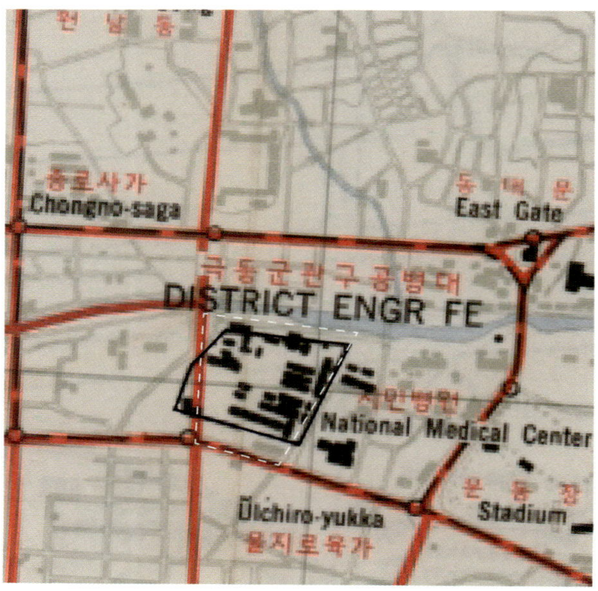

미 극동 공병단 기지, 1957 – 2018
U.S. Army Far East District, 1957 – 2018

서울시가지도, 미 육군극동기지국 제작 – 연도 확인불가
Seoul Road Map, U.S. Army Map Service Far East

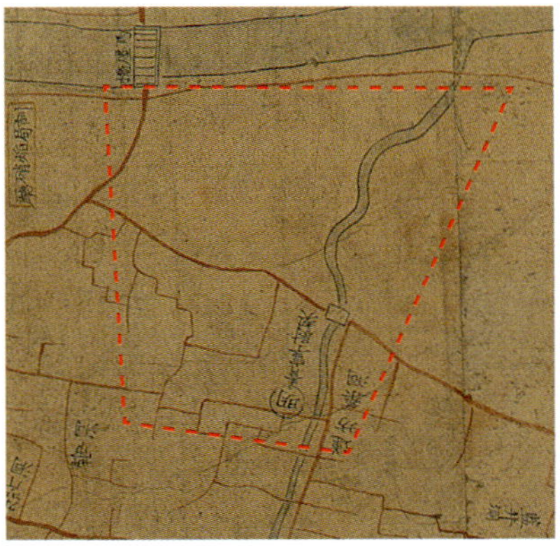

Late 18th Century

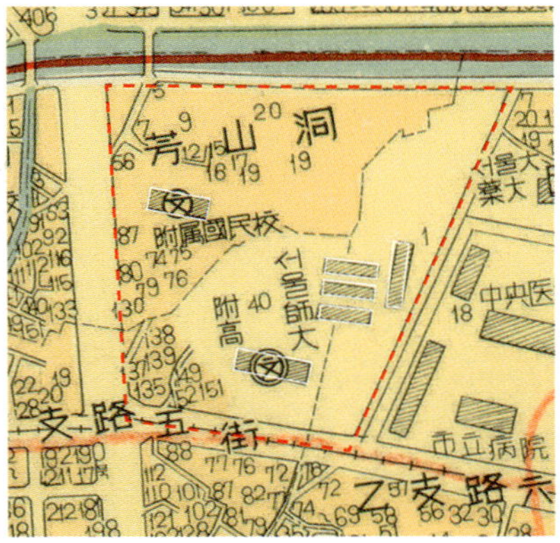

1958

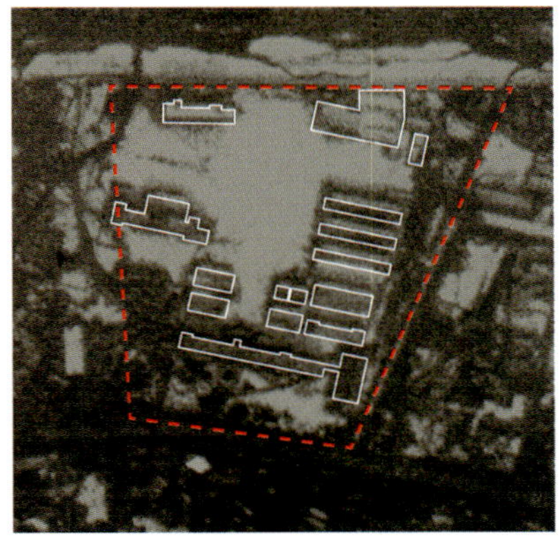

1947

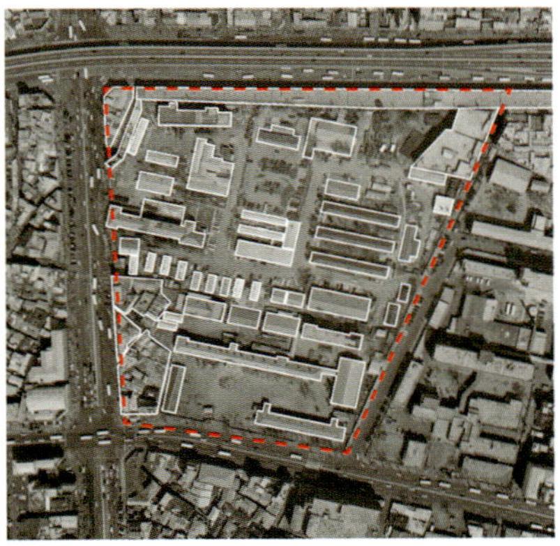

1975

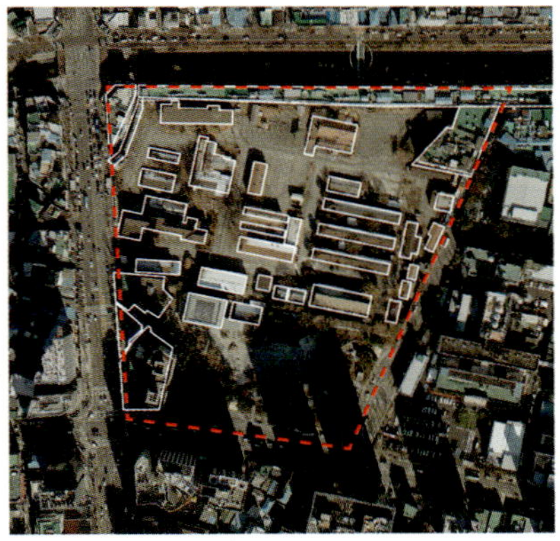

2020

Historically Built Area

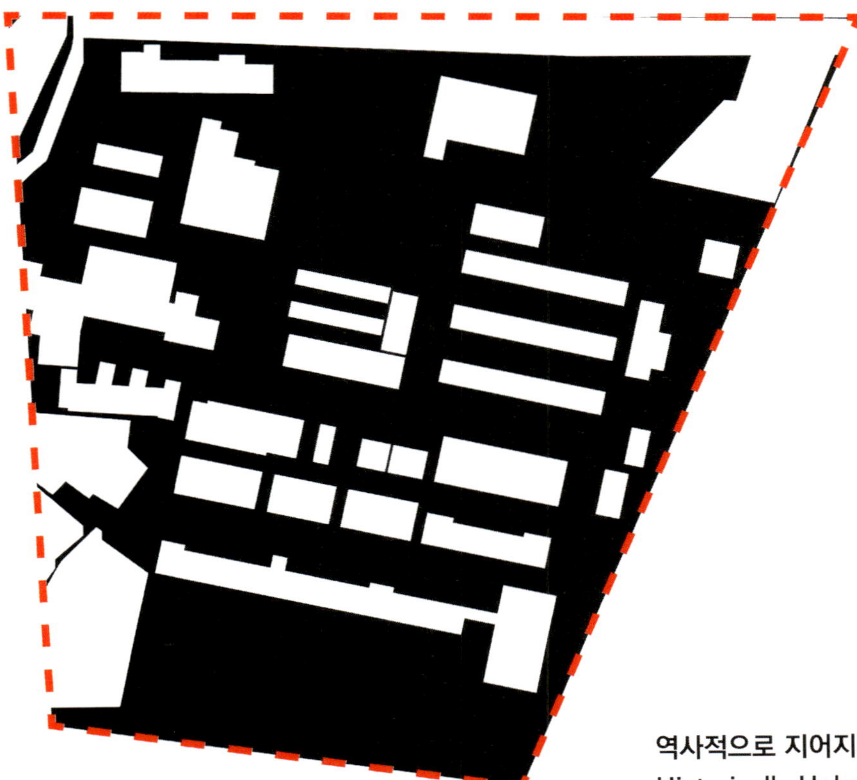

역사적으로 지어지지 않았던 땅
Historically Unbuilt Area

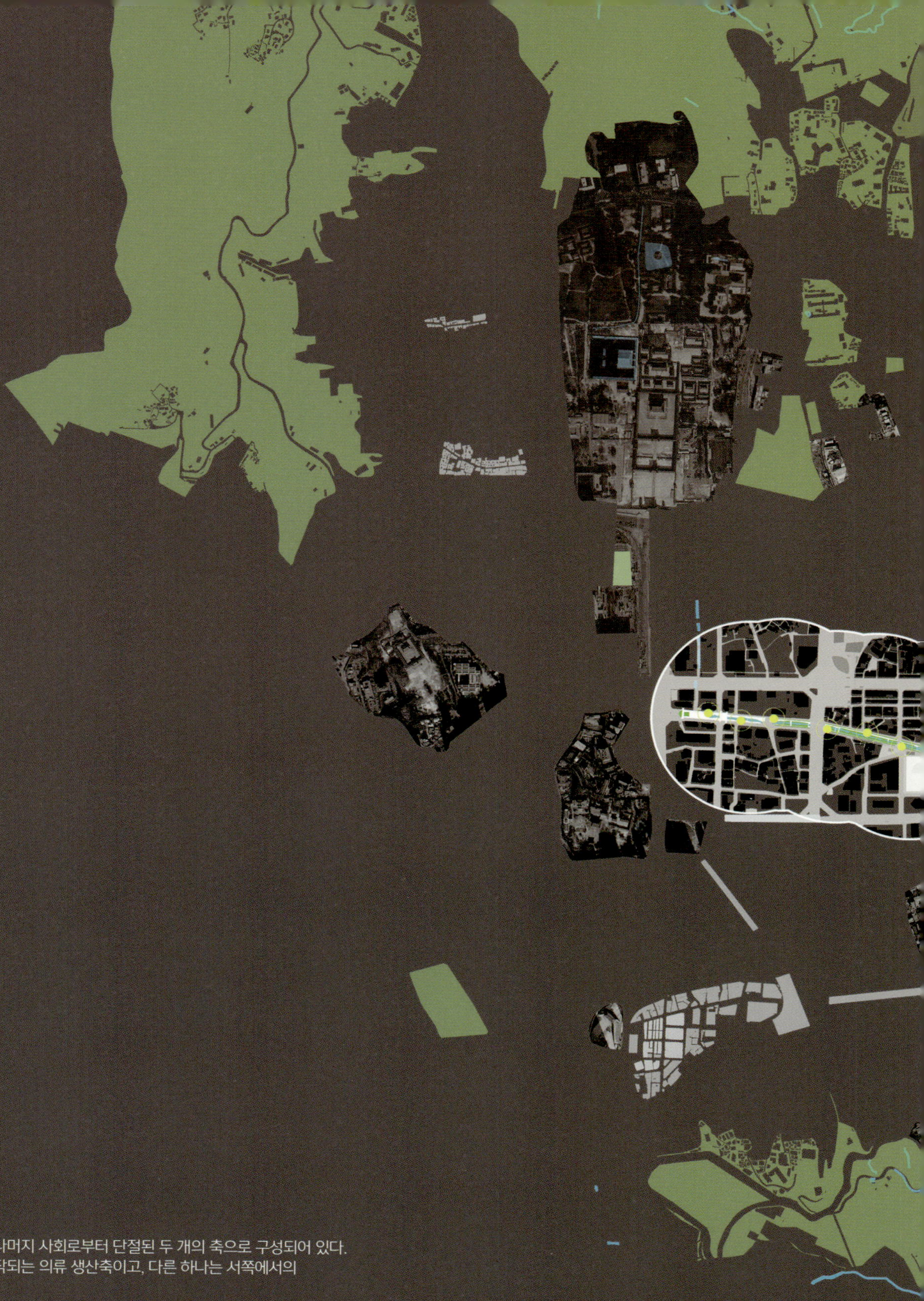

동대문 산업지구는 나머지 사회로부터 단절된 두 개의 축으로 구성되어 있다. 하나는 북쪽에서 시작되는 의류 생산축이고, 다른 하나는 서쪽에서의 기계 생산축이다.

The Dongdaemun industry is divided into two main axis, disconnected from the main society. One is the clothing manufacturing axis from the north, and the other is the machine-manufacturing axis from the west.

Site Analysis

Context study

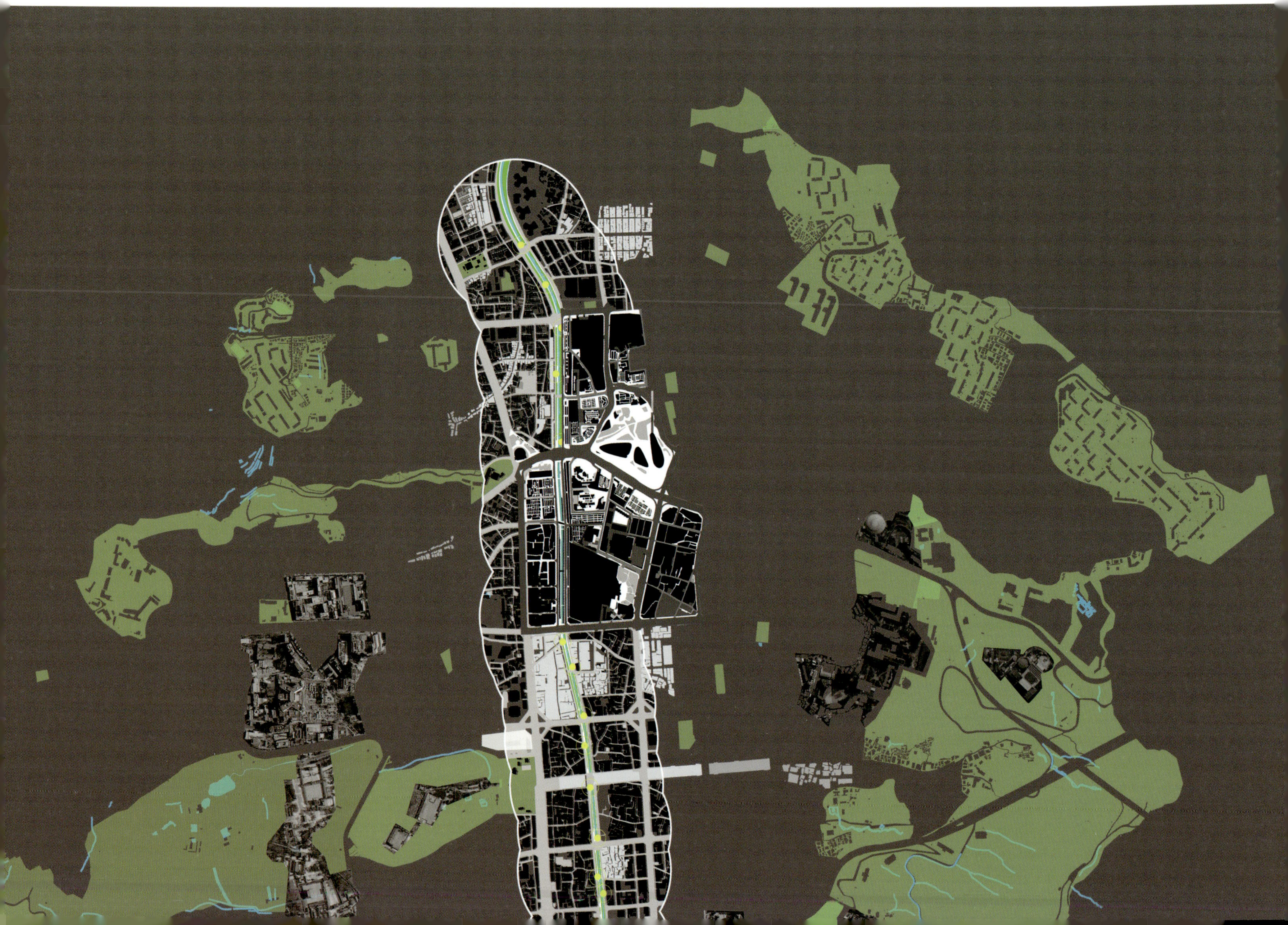

Conflict City, Seoul

Sangyup Yu, Juheon Kwon, Jinho Kim, Jeongwon Ko
Prof. Yerin Kang
Studio4-2, Seoul National University

Eun-Ma Apartment

There is various conflicts that hap-
pening at Eun-Ma apartment. Eun-Ma
was built in 1979 after GangNam area
Since then surrounding urban context
changing rapidly, but there was few inc
changes that happened in the apartmen
result, redevelopment is inevitable.

As redevelopment proposed lots
arised. First, there is conflict between
and Eun-Ma Complex, consent rate of

Cheonggye Highway

In terms of conflict, Seoul has historically had many conflicts between nature and cities.
In Seoul, there are places where the subject has changed from nature to city and from city to nature as hegemony changes as times change. And these were big events in the city that raising the conflict to the surface.

Bamseom Island exploded to build Yeouido, and Namsan Foreigners' Apartment was destroyed for Landscape of Namsan Mountain. And the Cheonggyecheon Stream has been changed shantytown to overpass, overpass to stream.

Cambell House in SagikDong

The Campbell Missionary House in Sagik-dong, registered as an architectural asset of excellence by Seoul City, exemplifies the conflict between the redevelopment association and Seoul City. In 2009, the Sagik 2 district was designated as an urban environment redevelopment area. Subsequently, approvals for the establishment of the association and project implementation were granted. In 2014, Seoul City introduced the "Comprehensive Conservation and Management Plan for the Fortress Villages surrounding Hanyangdoseong". Following this, the Seoul Urban Planning Committee revoked the urban environment redevelopment project of the Sagik 2 district association and canceled the association's establishment approval. In response, the association filed a lawsuit against Seoul City challenging the revocation and cancellation. In April 2019, the Supreme Court ruled in favor of the association, declaring the revocation and cancellation null and void. However, just five days after the victory, **Seoul City registered the 'Campbell Missionary House' within Sagik 2 district as an excellent architectural asset, using cultural heritage protection as a pretext to hinder redevelopment.**

Yeong

The Y
was establ
tion congr
6.6 squar
a deposit
Line in 20
those of ot
ects were p
zone prove
of Land. In
housing pr
deungpo Z

ment was upper than 75% in case of Eun-Ma apt residents but was less than 15% in case of Eun-Ma Complex. Second, there was conflict between apt and Seoul. Seoul government planned street that pass through apt complex, width of 15m, but apt refused to cooperated with that. Third, there is conflict between residents and owners who are not living. They have different plan and formed 2 different group.

These conflict again halt incremental changes... and apartment becomes more and more unsustainable.

Labor Condition in DDM area

The fashion industry in Dongdaemun, which has persisted since the 1960s, has endured in poor labor conditions. Jeon Tae-il, originally an assistant at Hanmi Company, became a cutter himself after the cutter was fired due to conflicts with the owner. In June 1969, he formed the labor movement group Babohoe at the Cheonggyecheon factory complex and later returned to Cheonggyecheon as a cutter for Wangseongsa, leading the labor movement. **In November, alongside a public display of the Labor Standards Act, he committed self-immolation at the entrance to the Peace Market by dousing himself in gasoline and igniting it with a lighter.** His death galvanized the formation of the Gyepebok labor union and revitalized the labor movement.

ng Po Slum District

ngpo Zzok-Bang Village, centered around guesthouses and brothels, he 1970s as urban poverty-stricken groups displaced by industrializa- **zzok-bang is a substandard residential space smaller than 's without a proper kitchen or bathroom, rented out without monthly or daily.** With the upcoming opening of the Sin Ansan ok-bangs within the construction zone have been demolished, but continue to be inhabited. Urban environmental improvement proj- rom 2015, but incorporating the brothel areas into the redevelopment iging. In January 2020, Seoul City, in collaboration with the Ministry ure and Transport and Yeongdeungpo District, announced a public mprove the living conditions and urban development of the Yeong- , Village.

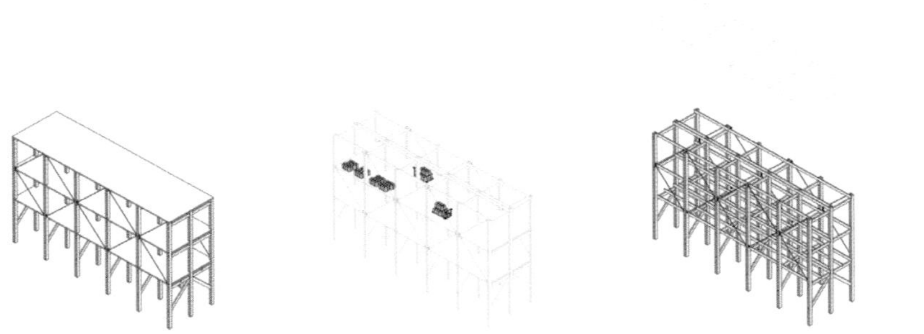

Modules-Bridge

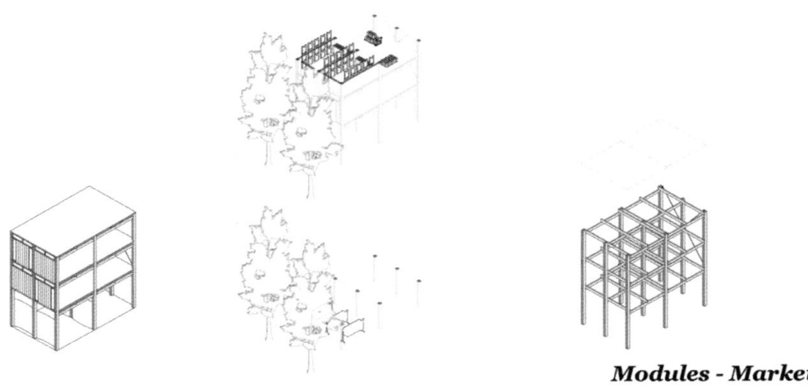

Modules - Market

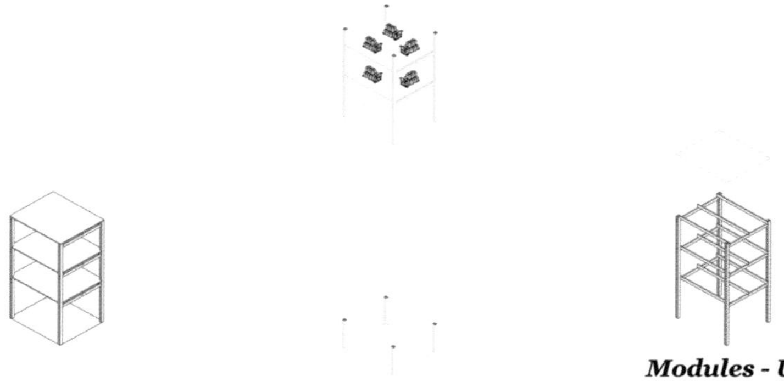

Modules - Universal

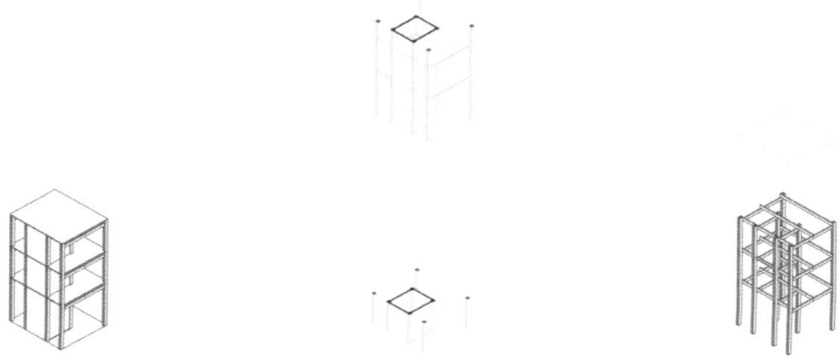

Modules - E/V

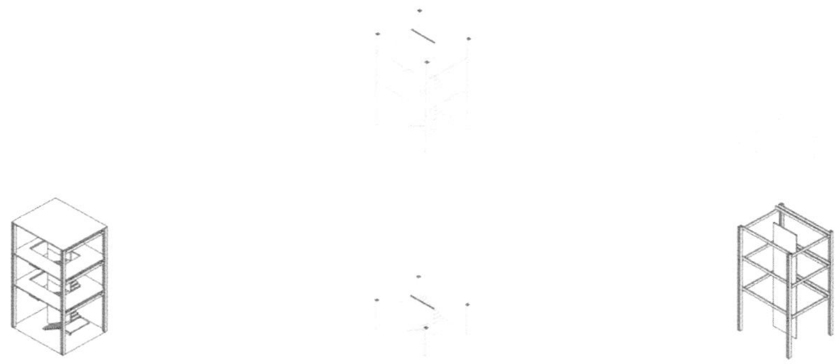

Modules-Stairs

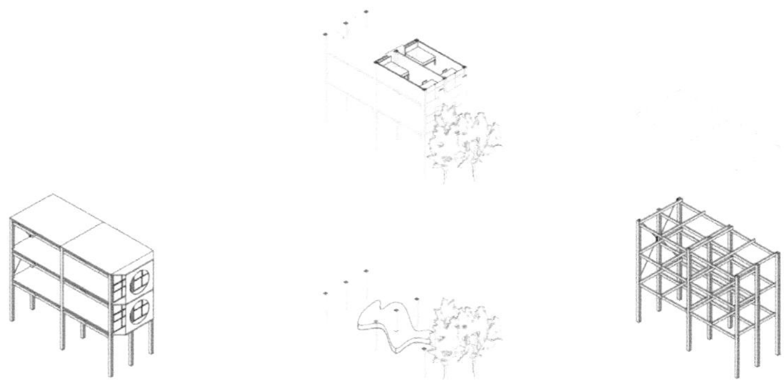

Modules - Resting

Modules - Public Stair

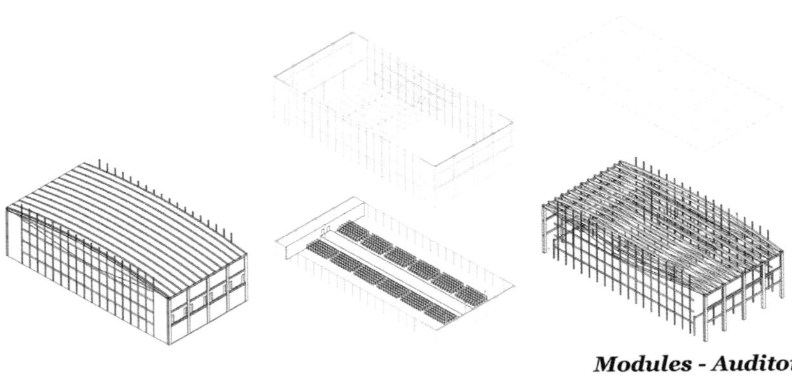

Modules - Auditorium

Organization01

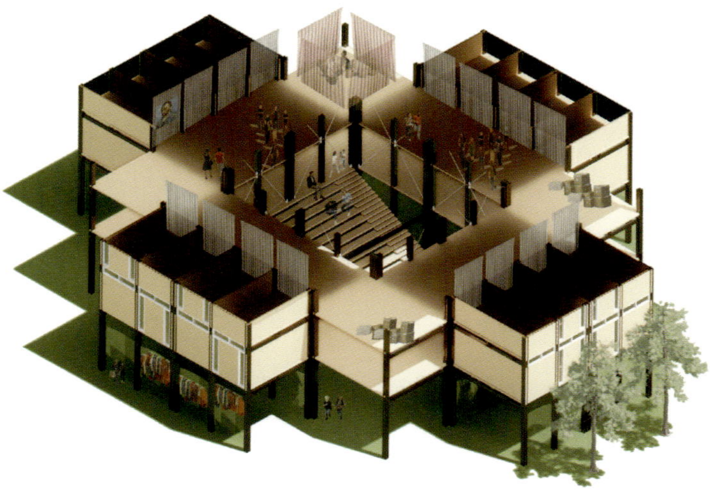

Organization01 F2 - Day

Organization01 F2 - Night

시-공간 그리고 공공-도매

Research 01 - 24시간 패션타운 (24H Fashion Town)

사이트인 미 극동반환부지(US Far-East District Camp Land)는 동대문 패션타운관광특구(DDM Fashion Town Tourist District)에 위치한 미지의 땅(Terra Incognita)입니다. 60년이 넘는 세월동안 미군 기지로 쓰이며 도시민의 출입을 금했습니다. 동대문 지역을 돌아다니며 24시간 작동하는 동대문을 확인할 수 있었고, 이를 표현하기 위해 24H 놀리맵을 제작해 보았습니다. 낮에 주로 열리는 패션 소매 클러스터와 밤에 열리는 패션 도매 클러스터간에 시공간적인 분리가 일어남을 확인할 수 있었습니다. 이는 사람들의 삶과 패션산업구조의 대비로 일어난 현상으로, 이런 시공간적 분리는 서로에게 방해를 주지 않기 때문에 편리한 것 같기도 하지만 서로가 다른 서로를 인식하지 못하기 때문에 앞으로 일어날 갈등의 씨앗이 될 수 있다고 생각했습니다.

Research 02 - 평화시장 (PyeongHwa Market)

사이트 북쪽에 위치한 평화시장은 사이트와 청계천 사이를 가로막고 있는 350m의 거대한 장벽처럼 느껴집니다. 이러한 상황에서 평화시장 주변에서 보행자와 다른 사용자들 간의 갈등을 확인할 수 있었습니다. 평화시장 북쪽의 큰 도로는 자전거전용도로 - 차도 - 주차장 - 인도 - 마켓 의 단면을 형성하고 있습니다. 하지만 좁은 인도마저 마켓의 상인들의 오버행(overhang)의 침범을 받고 있습니다. 또한 평화시장 내부의 24시간 열려있는 복도 또한 상점이 열려있을 때는 옷에서 나오는 화학물질로 인한 안좋은 공기가, 밤에는 복도 양쪽에 닫힌 셔터 때문에 일반 시민들이 이용하기 힘든 상태입니다. 평화시장은 동대문의 일반적인 도/소매 시장과는 다르게 의류 부자재와 의류 도매 판매가 동시에 일어나며 건물이 24시간 열려 있습니다. 하지만 열악한 보행환경과 도매 클러스터에서 이격되었기 때문에 생긴, 현 평화시장의 30%가 넘는 공실율과 낮은 매출은 현 상태가 개선이 필요하다는 점을 말하고 있습니다.

Design 01 - 평화시장의 전면성 전복 (Conversion of PH Facade)

평화시장의 북쪽 전면은 청계천의 복원 및 물류로 인한 보행환경의 악화로 더 이상 전면으로 사용될 수 없다고 생각했습니다. 평화시장의 남쪽 뒷면은 남쪽에 위치했던 미군 기지로 인해 아무런 역할을 하지 않다가 현재 새로운 가능성을 맞이하고 있습니다. 따라서 평화시장의 남쪽면으로 전면을 전복합니다. 전면성을 확보하기 위해 남쪽 입면을 구조만 남기고 허물어 개방합니다. 또한 평화시장 아래 청계천 level에서 저희 site로 들어올 수 있는 보행통로를 구축합니다. 청계천 이용객들은 이 통로를 통해 평화시장 남쪽 사이트로 진입합니다. 진입하면 큰 open space가 펼쳐집니다. 이 open space에서는 자연/그늘/휴식 공간이 제공됩니다.

평화시장의 전면이 바뀜에 따라 평화시장의 북쪽은 물류를 위한 공간으로 바뀌고 남쪽은 open space와 묶여 일반 방문자들이 사용할 수 있는 휴식 및 모임 공간이 됩니다. 이 두 성격의 공간은 새로운 지퍼 모양 경계에 의해 나눠집니다. 또한 평화 시장 1층 레벨은 도/소매 상점의 공간으로, 청계천 레벨(지하1층)은 방문자/녹지의 공간으로 크게 구분됩니다.

Design 02 - 확장 모듈 (Modular Expansion)

평화시장의 남쪽에서 평화시장은 확장됩니다. 평화 시장의 구조 그리드(5.7m*6.7m)를 그대로 확장시킨 구조물은 기존 평화시장의 철근 콘크리트 구조와는 다르게 가벼운 철골/목 구조로 되어 있습니다. 이 건식 구조와 그리드 체계의 성질로 인해 남쪽 부분은 확장성(가변성)을 가집니다. 건축가는 모듈과 구성 규칙을 제안하지만, 이 그리드 형식의 구조물은 남쪽 공원으로 시간의 흐름과 사용자의 요구에 대응하여 조립/분해 될 수 있습니다. 이를 위한 몇 가지 module을 제안했습니다. 1층과 2층 레벨에 서는 도/소매 상점과 상점 사장님들이 쉴 수 있는 공간, universal space가 구축됩니다. 지하 레벨은 display/체험의 공간입니다. 근처 대형 쇼핑몰에서도 쇼핑 몰 앞 공개공지에서 천 막을 치고 옷을 파는 것처럼 사람들은 위에 도매 시장에서 판매되는 옷을 입어보고 구매할 수 있습니다. 이 옷은 (마치 평화시장이 세워졌을 때 상인들의 연합으로 세워졌듯이) 평화시장을 총괄하는 새로운 사업체의 등장으로 가능해질 것입니다. 이 사업체는 도매시장에서 옷을 구매해서 아래 display장소에 편십샵의 느낌으로 판매합니다.

Design 03 - 모듈의 조합(Organization of Modules)

여러 모듈은 특정 원리에 입각하여 조합됩니다. 첫 번째는 환기(ventilation)입니다. 기존 평화시장은 중복도 형식으로 되어 있어 환기가 제대로 되지 않았습니다. 새로운 마켓 모듈은 앞 뒤로 오픈된 공간이 있어야 하며 입면에는 통풍창이 2개 이 상 뚫려 있어야 합니다. 기존에 매우 안좋은 공조 환경에 있어야 했던 상인들은 새로운 모듈에서는 새 옷과 부자재에서 나오는 유해 물질로 인한 걱정이 줄어들 것 입니다. 두번째는 Universal space 입니다. 이 공간은 낮-공간(공공)과 밤-공간(도 매)사이에 존재하며, 낮에는 낮-공간에 포함되어([낮- U] - 밤), 밤에는 밤-공간에 포함되어 기능합니다(낮- [U - 밤]).이 공간으로 인해 새로운 도매 시장은 공공과 섞여, 공간이 효율적으로 쓰일 수 있을 뿐더러, 새로운 판매 구조의 가능성을 모색할 수 있습니다. 나아가 패션 산업구조 자체의 변화가 생기더라도, 즉 이 공간이 더 이상 전통적인 도매 시장 뿐만이 아니라 새로운 형태와 다른 물건을 파는 판매의 공간이 되더라도 제대로 작동할 수 있을 것입니다.

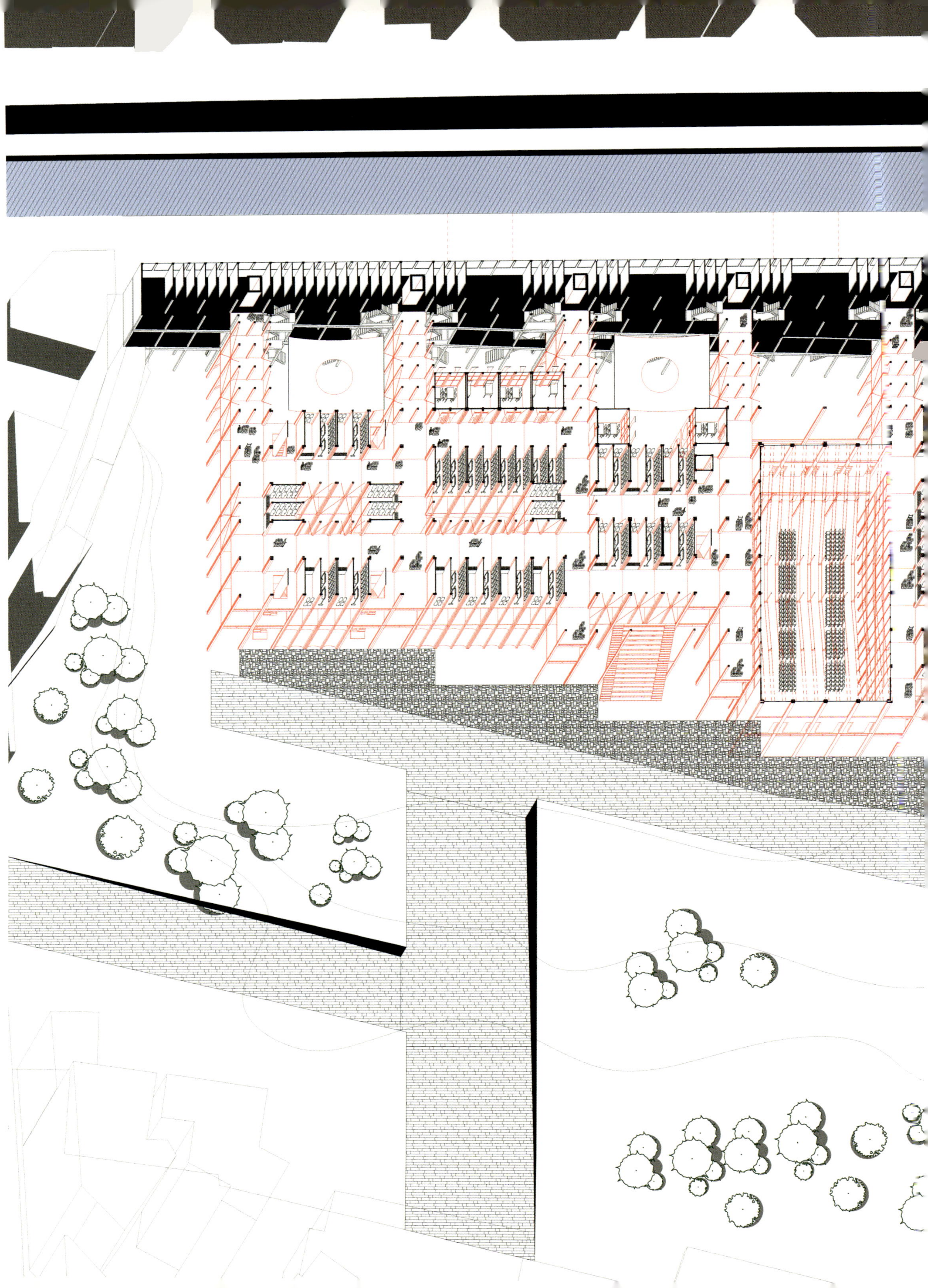

0
10

0
10m

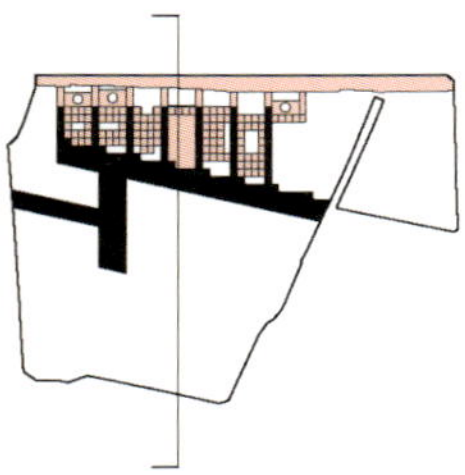

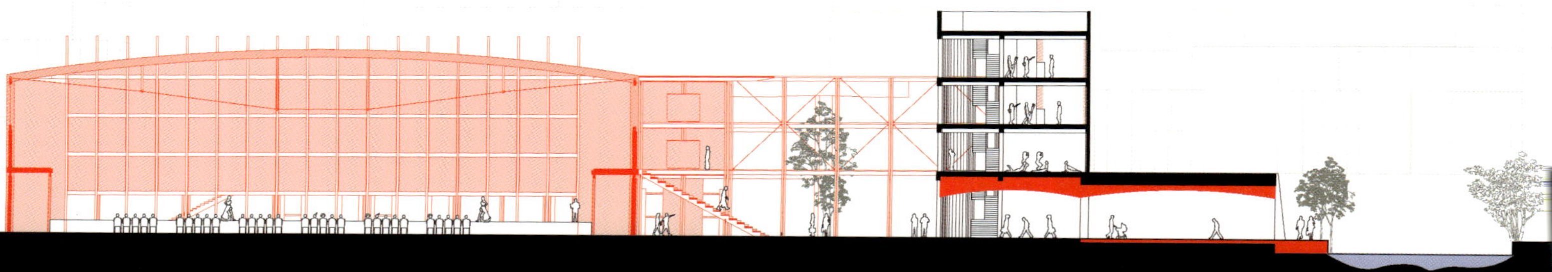

Interweaving Time-Space & Public-Wholesale

Research 01 - 24H Fashion Town

The US Far-East District Camp Land, situated within the DDM Fashion Town Tourist District in Dongdaemun, has long remained an enigmatic Terra Incognita. Its history as a military base for over six decades barred access to urban civilians. Exploring the Dongdaemun area revealed a distinctive operational pattern - a 24-hour cycle within the fashion town. Through the creation of a 24H Nolli map, it became evident that spatial and temporal segmentation occurred between the daytime retail fashion cluster and the nighttime wholesale fashion cluster. This division, arising from the contrast between people's lifestyles and the structure of the fashion industry, seemed convenient due to the lack of obstruction between these realms. However, the failure to recognize each other's existence may potentially sow seeds of future conflict.

Research 02 - PyeongHwa Market

PyeongHwa Market, positioned to the north of the site, stands as a colossal 350-meter barrier between the site and the Cheonggyecheon. Observations around PyeongHwa Market unveiled conflicts between pedestrians and various other users. The major road north of the market delineates a sequence: bicycle lane, roadway, parking, sidewalk, and market. Even the limited sidewalk space encounters encroachments from market vendors. Additionally, the market's 24-hour open corridor suffers from air quality issues during business hours due to chemicals emanating from clothing items, while at night, shuttered entrances deter public access. Differing from conventional Dongdaemun retail markets, PyeongHwa Market engages in both clothing materials and clothing wholesale. Nevertheless, its current state, marked by over 30% vacancy and diminished sales, indicates a pressing need for improvement.

Design 01 - Conversion of PH facade

Acknowledging the north facade's impracticality owing to deteriorated pedestrian conditions resulting from Cheonggyecheon restoration and logistics, a proposal emerges to overturn the PyeongHwa Market's south-facing elevation. Preserving the structural integrity, the south-facing frontage will undergo demolition, enabling access from the southern direction and facilitating a new underground pedestrian pathway from the Cheonggyecheon level to our site. Upon entry, visitors will encounter a spacious open area designed to provide natural elements, shaded zones, and relaxation spots.

This alteration relegates the north side to a logistics space, while the transformed south side becomes an accessible relaxation and gathering space for the general public, delineated by a newly proposed zipper-shaped boundary. The first floor accommodates retail shops, while the Cheonggyecheon level (basement) is slated for visitors and green spaces.

Design 02 - Modular Expansion

PyeongHwa Market's expansion from the southern side involves expanding the market's structural grid (5.7m*6.7m) using a lightweight steel and wood structure, departing from the current fibered concrete framework. This modular structure, based on a grid system, grants flexibility (variability) to the southern part. While architects propose modules and construction rules, this grid-based structure allows for assembly and disassembly according to time and user needs. Several modules have been proposed: retail spaces on the first and second levels and a universal space, and a resting area for shop owners. The underground level serves as a space for displays and experiences, where people can try and purchase clothes potentially led by a new overseeing entity(like merchants association initially established PyeongHwa Market,). This entity procures clothes from wholesale markets and retails them in the display area akin to a edit shop.

Design 03 - Organization of Modules

Modules are combined following specific principles. Firstly, ventilation is crucial, necessitating open spaces and ventilation windows aside the facade. Existing market areas with double corridor structure suffered from poor air; the new modules aim to alleviate concerns about harmful substances emitted from new clothing and materials. Secondly, the concept of a universal space is introduced. This space exists between day (public) and night (wholesale) functionalities, operating as part of the day-space ([Day- U] - Night) during the day and part of the night-space (Day - [U - Night]) during nighttime. This adaptable space not only efficiently utilizes public areas but also explores the potential for new selling structures. It accommodates potential shifts in the fashion industry's structural dynamics, serving as a functional space even beyond traditional wholesale markets.

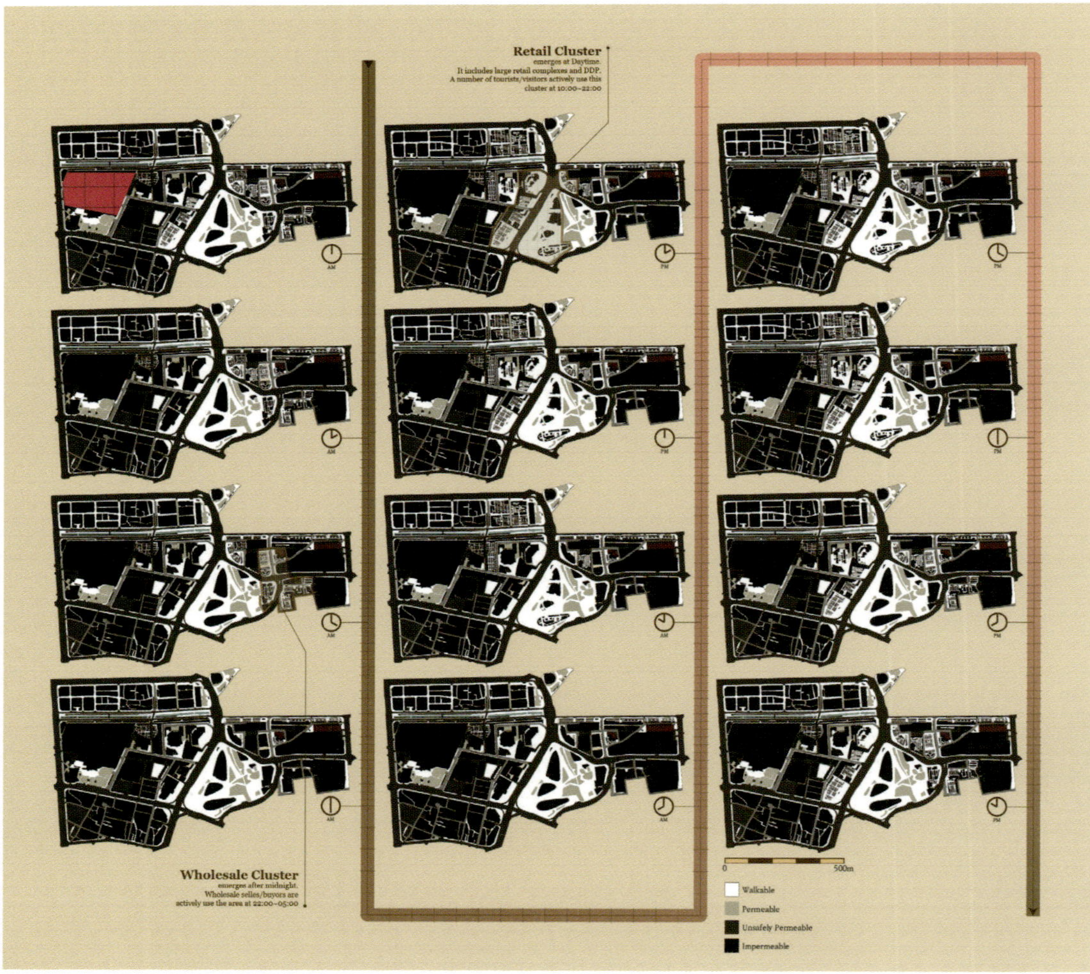

Conflict Potential Between Retail Occupants and Wholesale Occupants

We Consider DDM Fashion district area as a place in which possible conflict between retail seller/buyor and wholesale seller/buyor can happen. As mentioned before, conflicts can happen when the distance between two clusters are too far not to mention the case of 'too close'.

Retail occupants and wholesale occupants can't even recognize each other since they are separated in time and space. Then opposing forces can take place. For example, the spaces that are used at night time are closed in daytime and not opened to pedestrain, and vice versa. When we move our sight from an urban perspective to more social perspective, there might be another conflicts. These two occupants groups are likely to opposing each other in decision making concerning DDM district. Because they usually can't recognize each other and this blocked sight makes them hard to understand each other, and regard each other as part of the bigger community groups that share the same spaces.

24H NolliMap

DDM fashion district area is categorized by the level of public openness to pedestrians. It has 4 categories which are 'Walkable', 'Permeable', 'Unsafely Permeable', 'Impermeable', and are getting darker from former to latter.

We recognized shopping complexeswhich has semi-public corridors and halls inside the building as a public area, since they represent the urban condition inside themselves.

Furthermore, to examine DDM fashion district area more accurately, we made Nollimap in 24-Hour basis. As time passes in a day one can notice shift of the open spaces in the area as the urban shopping complexes open/close.

Conflict Between Pedestrians And Other Occupants

The picture below shows the condition of street around the site. The one in the center is Nollimap in which spaces are categorized by permeability(of pedestrains). The spaces around the site all in all has bad condition in terms of walkability. And some areas are not permeable and form cull-de-sac in return. While large retail cluster that contains DDP and large shopping complex are highly walkable and attracts large amount of visitors(or pedestrains), the streets around the site are less walkable.

We redefine this condition as conflict between pedestrains and other occupants. For instance, street in front of the Pyeong-Wha market is invaded by stores along it. Market owners deploy their stores to the streets which should be used publicly.

In response, pedestrians are reluctant to use those streets. And vicious circle take place as they exclude another.

A

Pedestrians vs Stores & Motorvehicle

Stores deploy themselves outside of the building, to the facing street. It intrude the street that belongs to public in first place.
Moreover, motorvehicle that are used to transport products threats pedestrians. And those are parked along to street.

Pedestrians vs BDG

These group of building made cul-de-sac and it makes PH market acts as a long barrior, 350m long. cul-de-sac makes physically/cognitively close space far in travel distance. Then it becomes kind of urban labyrinth.

D

B

Pedestrians vs PM/Logistics

Street at the left side of the photo is a bike lane that people can't walk. On the right, trucks and vans for logistics intrude sidewalk. Even, cranes that transport product from ground level to rooftop are on operation in daytime.
Outworn overhangs make pedestrians detour the street.

E

Pedestrians vs Stores/Motorvehicle

C

Pedestrians vs Closed Store

The photo is takened in the PyeongWha market arund 17:00. Even corridor of PW market act as semi-public street, it is no longer used as street when the stores along it is closed.
Bad air condition and narrowness of the corridor also are the negative factors.

G

Pedestrians vs Store/Motorvehicle

F

Pedestrians vs Skateboarders

Hullyeonwon Park is popular to skateboarders and physical conditions such as skidding lane are also suitable for skateboarders.

Date of Incident	Location	Category
10 AM, 10th, Mar, 2014	Bangsan Market	Disaster
Name of Incident		
Bangsan Market Fire Accident		

Date of Incident	Location
13th, July, 1965	Ch
Name of Incident	
Displacement from Cheonggye Overpass Construction	

Date of Incident	Location	Category
12:53PM, 25th, Jul, 2003	USA Army Corps of Engineers Far East District	Protest, Ideology
Name of Incident		
Hanchongryun Protest against US Army Presence		

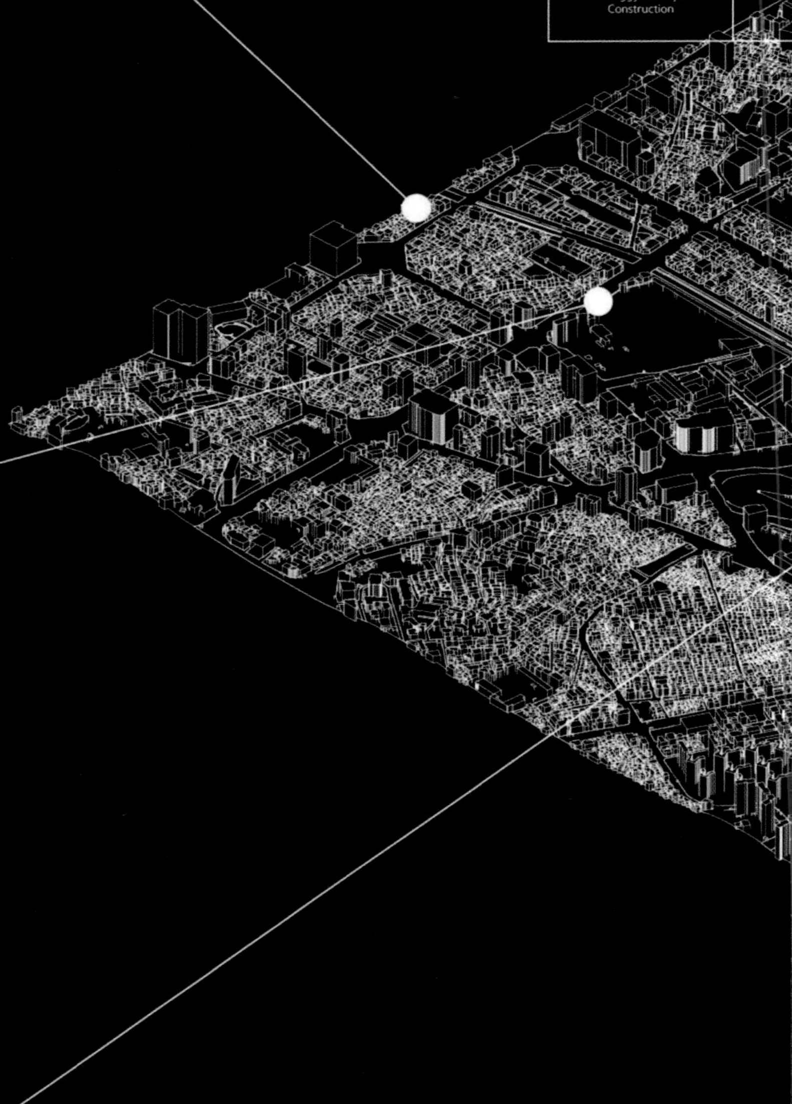

Date of Incident	Location	Category
12:39AM, 22nd, Sep, 2019	Jeil Pyunghwa Market	Disaster
Name of Incident		
Jeil Pyunghwa Market Fire Accident		

Dongdaen

Tracinc

Category
Demolition, Displacement

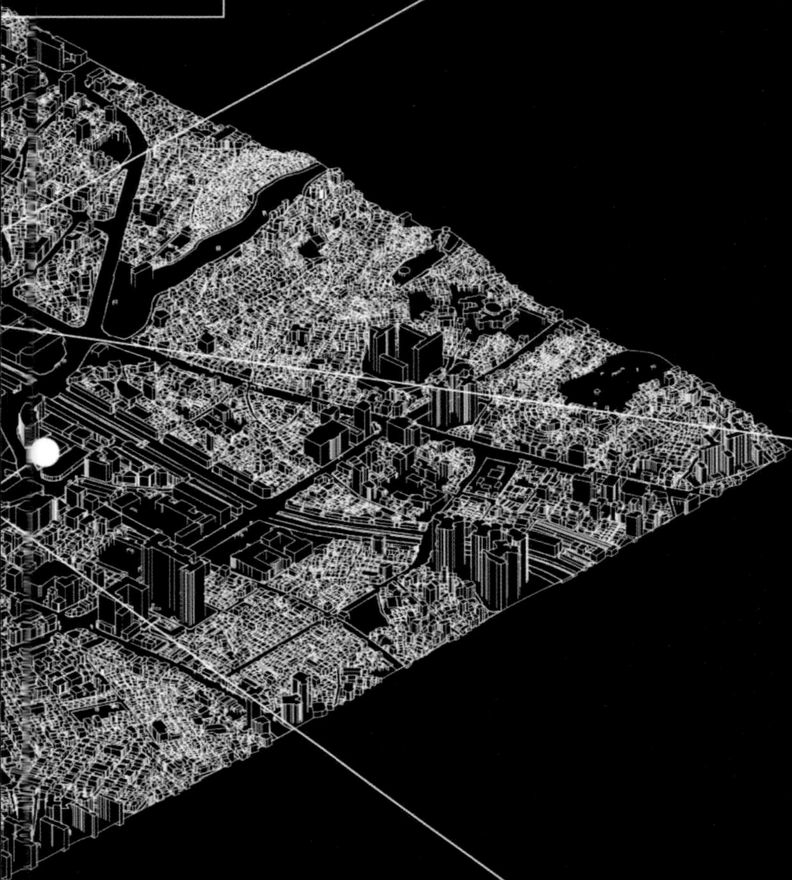

Date of Incident	Location	Category
10:57 PM, 20th, Oct, 2014	Dongyang Fiber Corp.	Disaster
Name of Incident		
Dongyang Fiber Corp. Fire Accident		

Date of Incident	Location	Category
13th, Nov, 1970	Pyeonghwa Market	Protest, Labor Rights
Name of Incident		
Jeon Tae Il Self Immolation		

Date of Incident	Location	Category
17th, Apr, 2008	Dongdaemun Stadium	Demoliton, Displacement
Name of Incident		
Pungmul Market Demolition		

n Casefile

ncidents

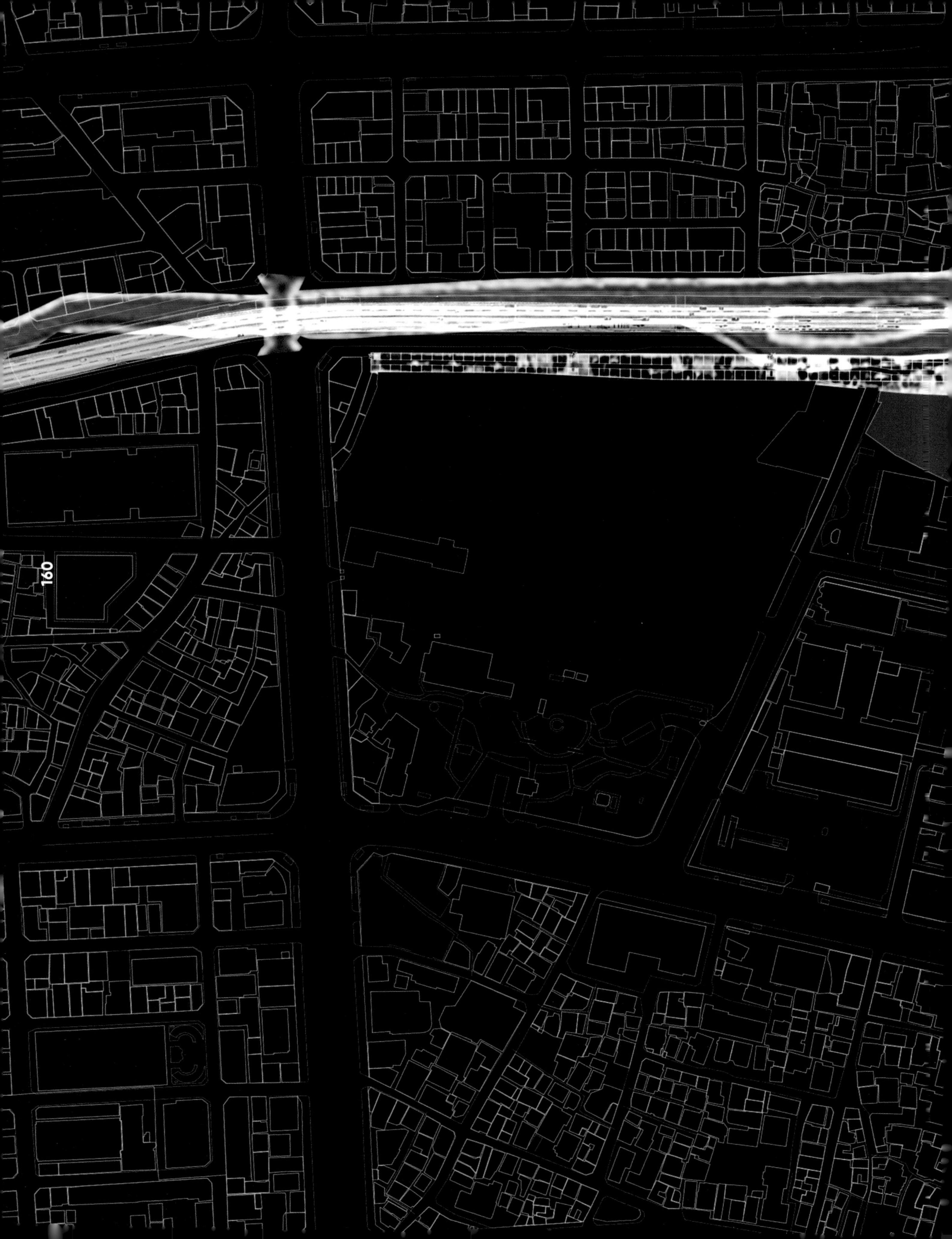

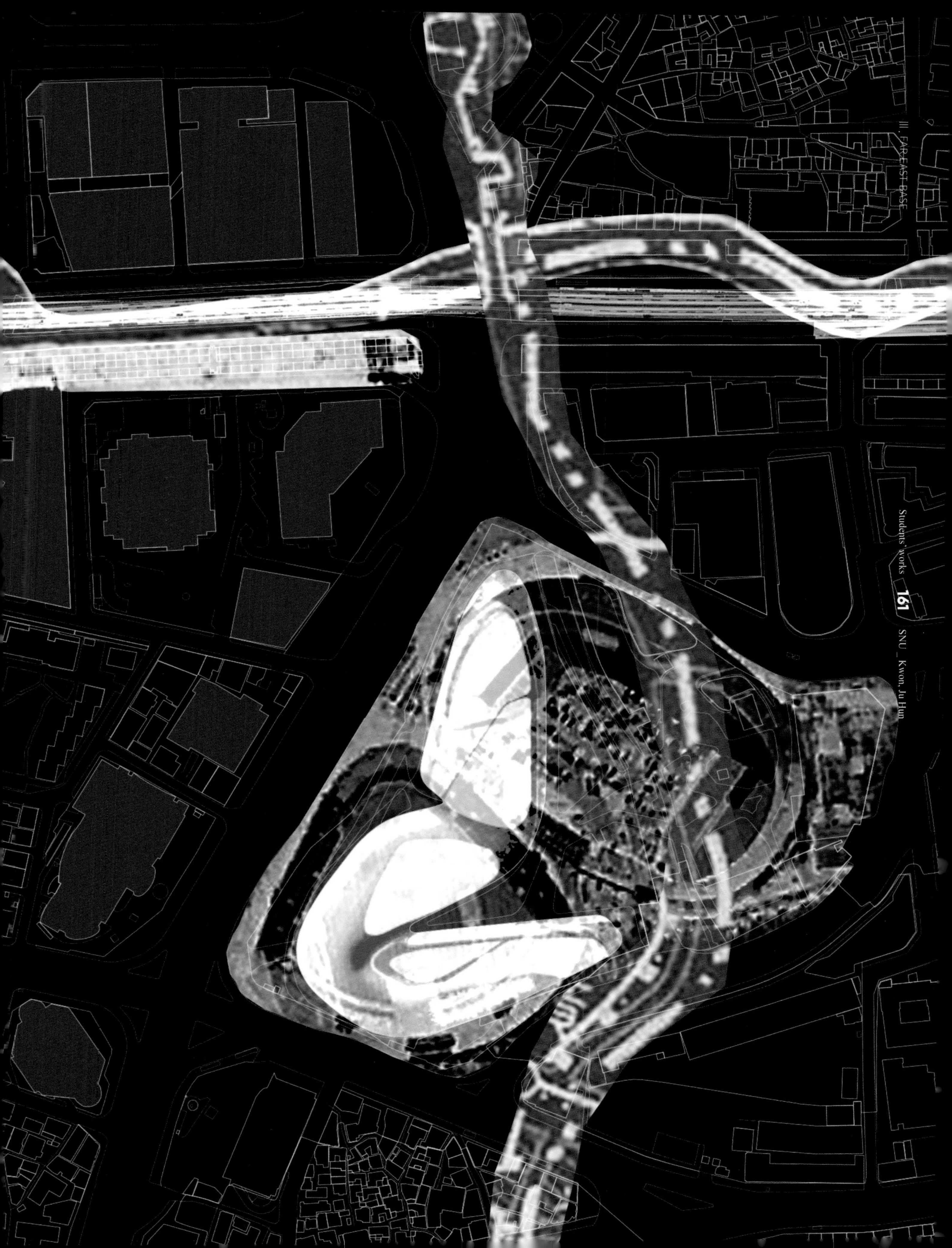

Ground Plan (+0)

팀 리서치와 더불어 팀 내에서 주목했던 서울의 어바니즘은 갈등도시로서 서울이 갖는 정체성이었다. 그 중 본 설계에서 알아보고자 했던 것은 미극동 공병단 부지라는 이 땅 위, 혹은 그 주변부로 있어왔던 “도시에 대한 권리”에 대한 투쟁, 그리고 그로 인한 소요 사태다. 동대문 지역은 여러 양상의 ”도시에 대한 권리“와 관련된 소요 사태를 경험해왔다. 일제강점기, 1955년, 1958년, 1977년 등에 있었던 여러차례의 청계천 복개공사, 1967년부터 1971년까지 진행되었던 삼일고가도로 공사는 청계천변에 있어왔던 많은 판잣집들을 철거, 주민들을 봉천동, 신림동, 상계동, 광주대단지 등의 타지역으로 강제 이주시키는 결과를 초래했었다. 1930년대 건립된 경성운동장 은 조선시대 사대문 안을 규정하던 한양도성이라는 중요한 문화재를 허물고 지은 결과물이며, 동대문운동장으로 이름이 바뀌고 나서 DDP건설로 인해 철거될 때까지 주변 상가로부터 수용했던 상인들을 시 차원에서 강제이주시켰던 장소였다. 1960-70년대 전태일 분신자살사건을 필두로 한 평화시장 노동자들의 근로기준법 준수 요구가 있었으며, 이후로는 청계피복노조가 전태일의 정신을 이어 받아 노동운동과 민주화 운동을 진행해왔다. 또한 2003년 7월 한총련 소속 대학생 20명이 미군 주둔과 한반도 전쟁 위협에 반대하는 기습 시위를 하는 등 동대문 지역은 생존권, 노동권, 이데올로기가 끊임없이 국가, 외세 등 거대한 권력에 의해 침해당하고, 이에 대한 저항을 하며 “도시에 대한 권리”를 회복하고자 투쟁하는 장소로서 역할을 해왔다.

해당 프로젝트는 이러한 동대문 지역의 장소성을 반영하여 “도시에 대한 권리”를 회복하고 역사 속에 있어왔던 “도시에 대한 권리”가 침해되고 이를 회복하고자 벌어졌던 투쟁을 기억할 수 있는 공원을 조성하고자 한다. 사이트 북쪽의 청계천과 평화시장, 사이트 남쪽의 동대문 패션벤처타운을 지하로 연결하고, 이로 인해 조성된 선형의 동선은 휘발되어 없어진 사건들을 Transcribe한 폴리들을 지나며 동대문 지역의 사건들에 대한 기억을 환기시킨다. 동선의 최종 장소인 공원은 누군가에게는 일상으로부터 벗어날 수 있는 공간임과 동시에 누군가에게는 자신의 혹은 공동체의 ”도시에 대한 권리“를 주장할 수 있는 장소로서 역할을 할 수 있기를 기대하며 조성하였다.

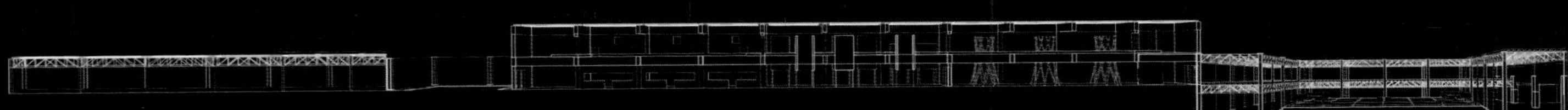

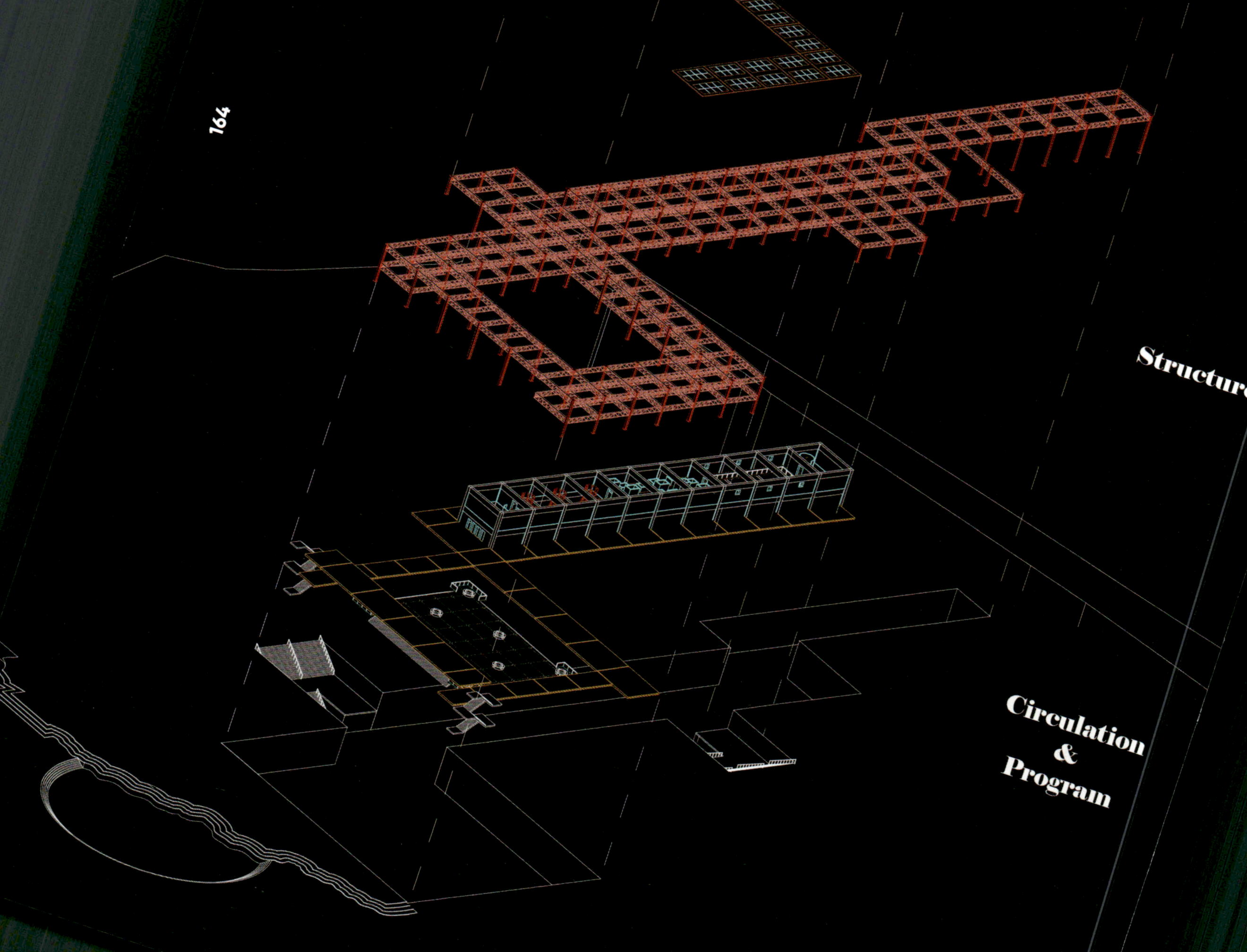
Structure
Circulation & Program

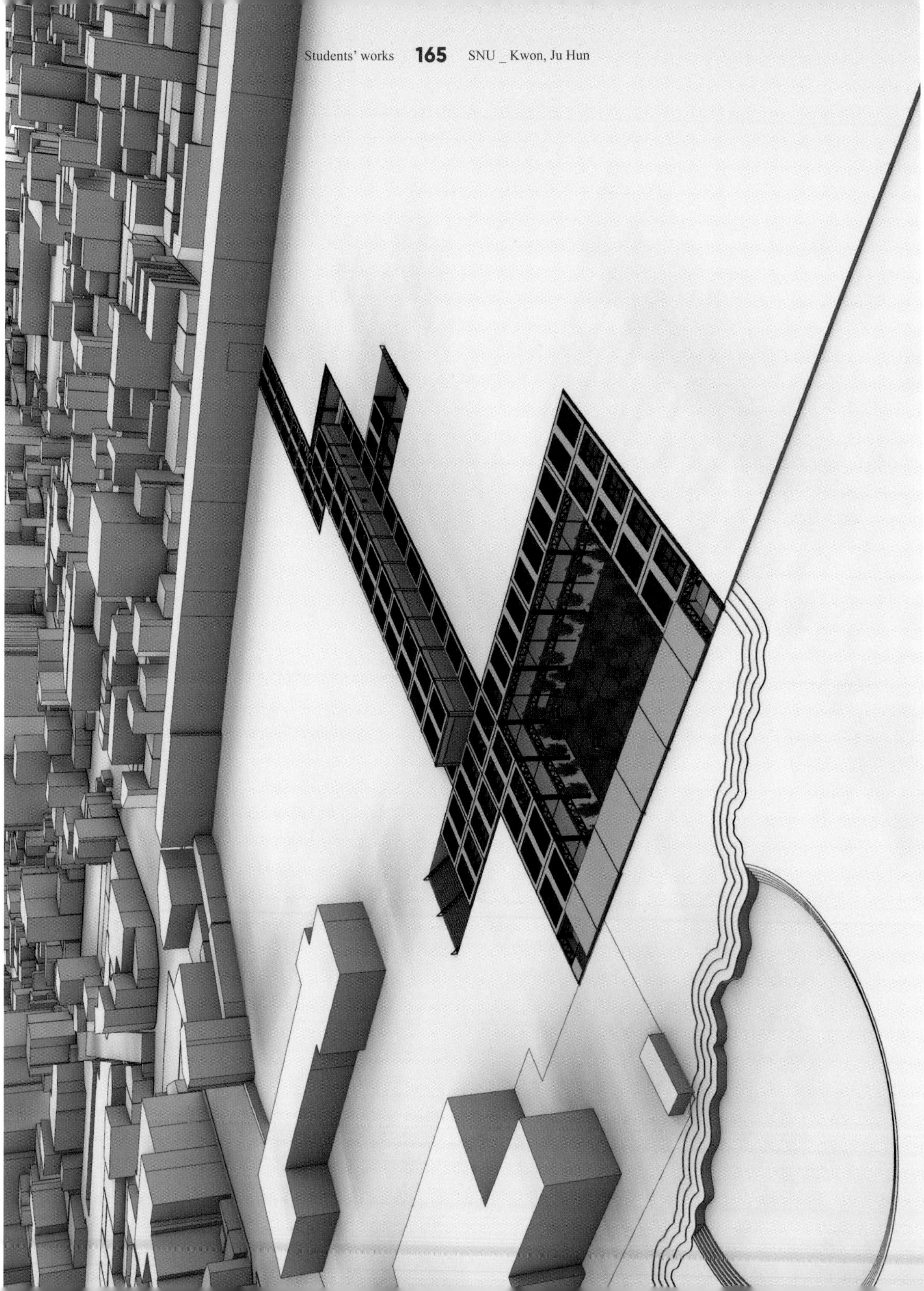

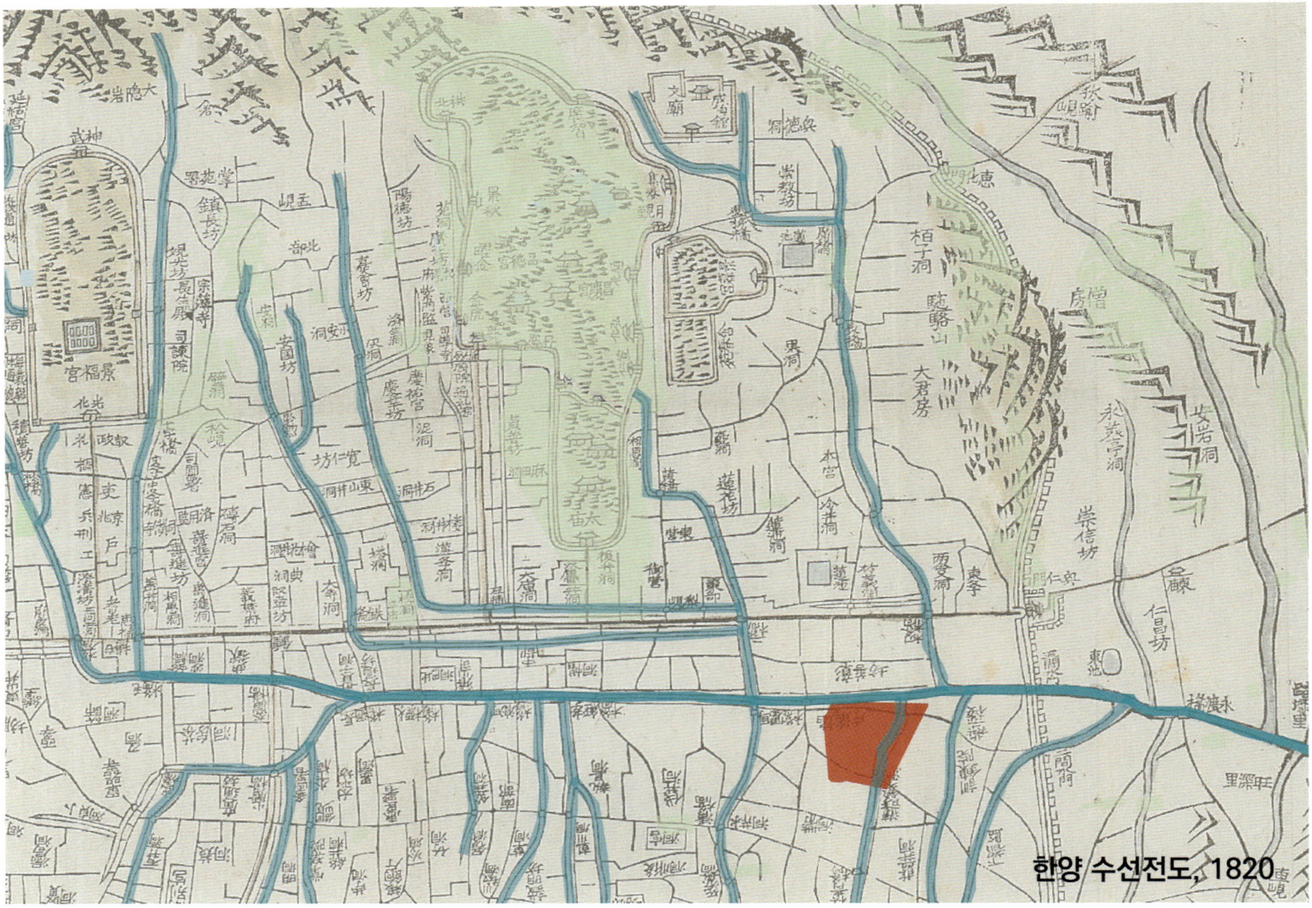

한양 수선전도, 1820

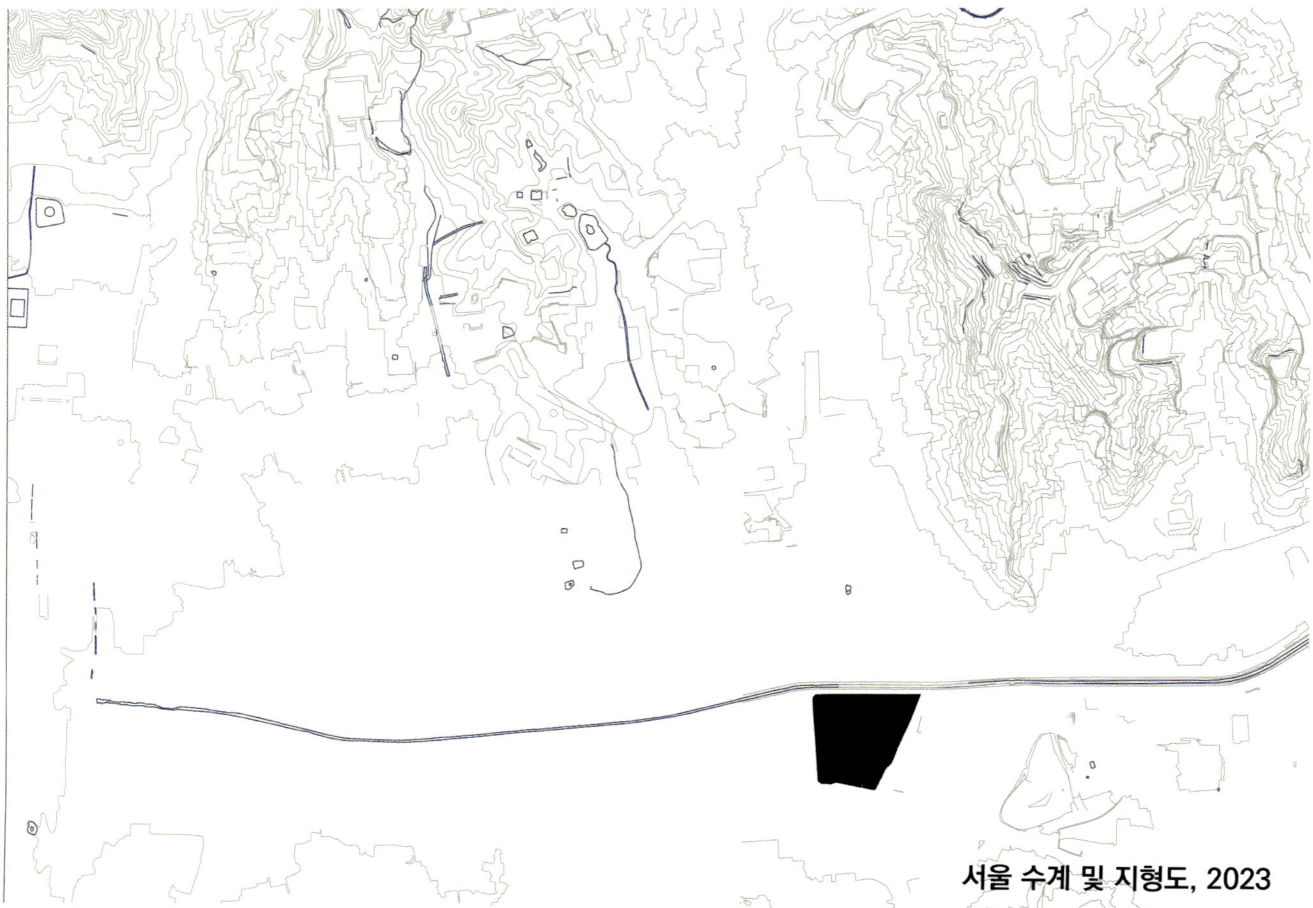

서울 수계 및 지형도, 2023

재자연화 중인 청계천 (Rewilding)

청계천과 지하상가(공간)

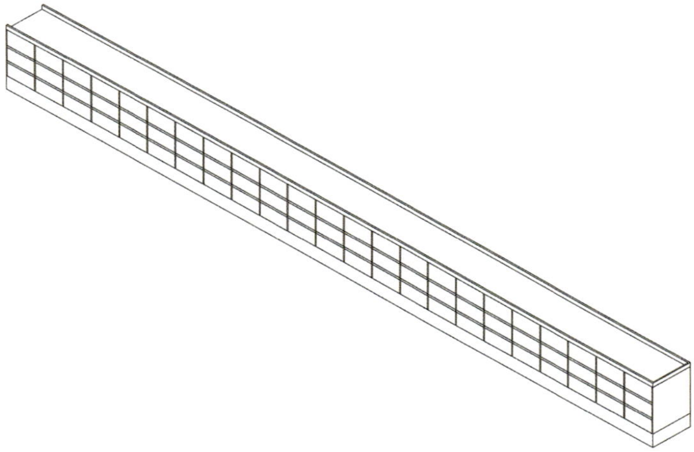

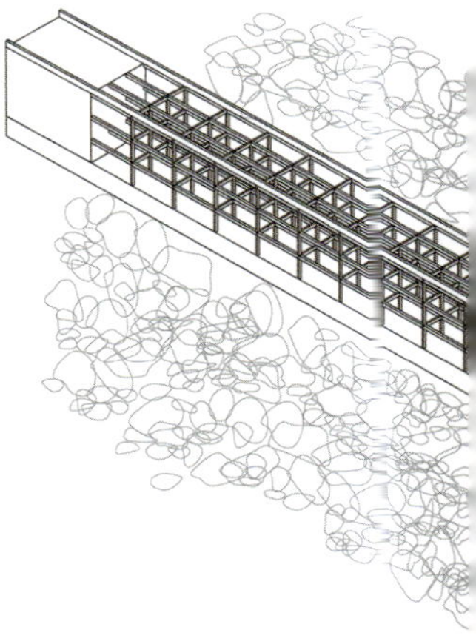

기존 평화시장

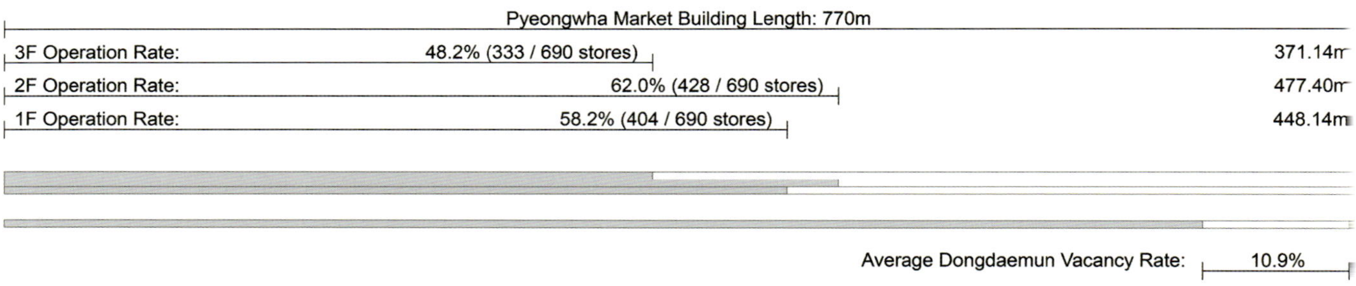

평화시장 운영율과 동대문 공실률

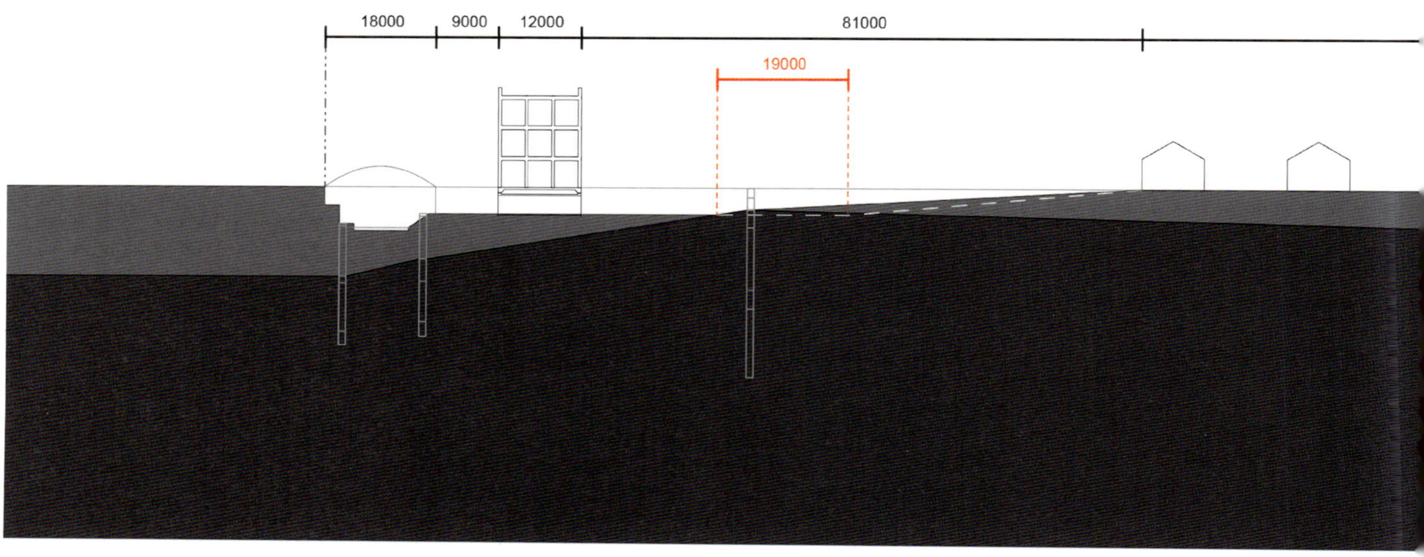

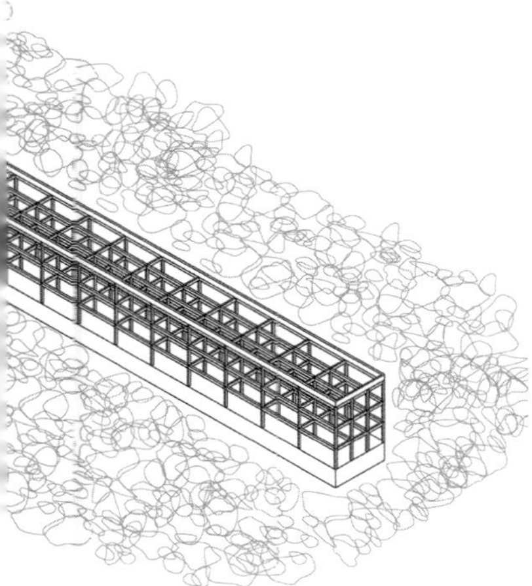

폭파된 평화시장과 건설폐기물

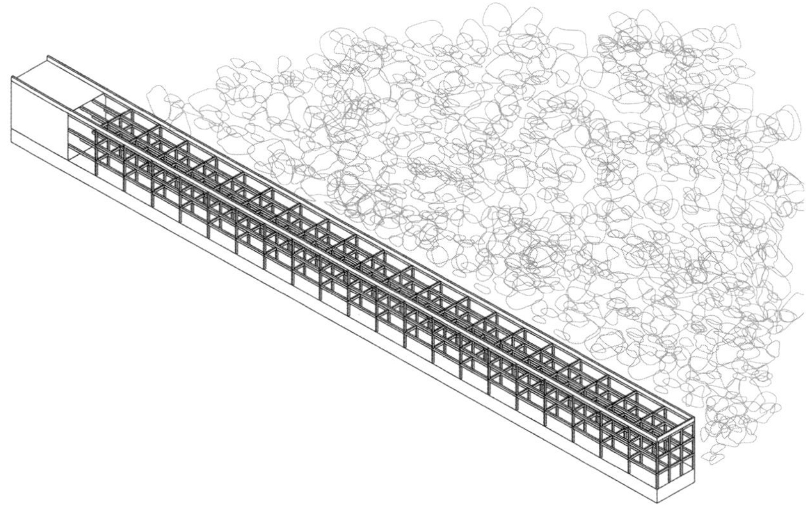

Mud-scape을 위한 자연제방으로 재활용

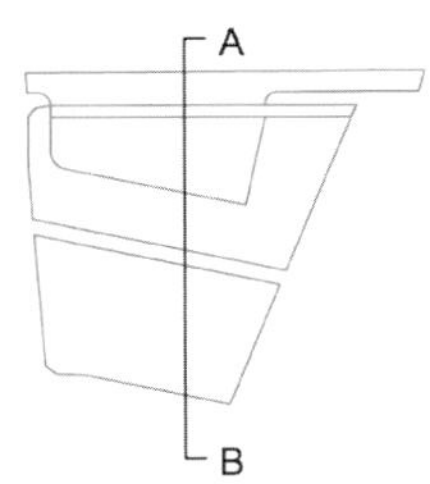

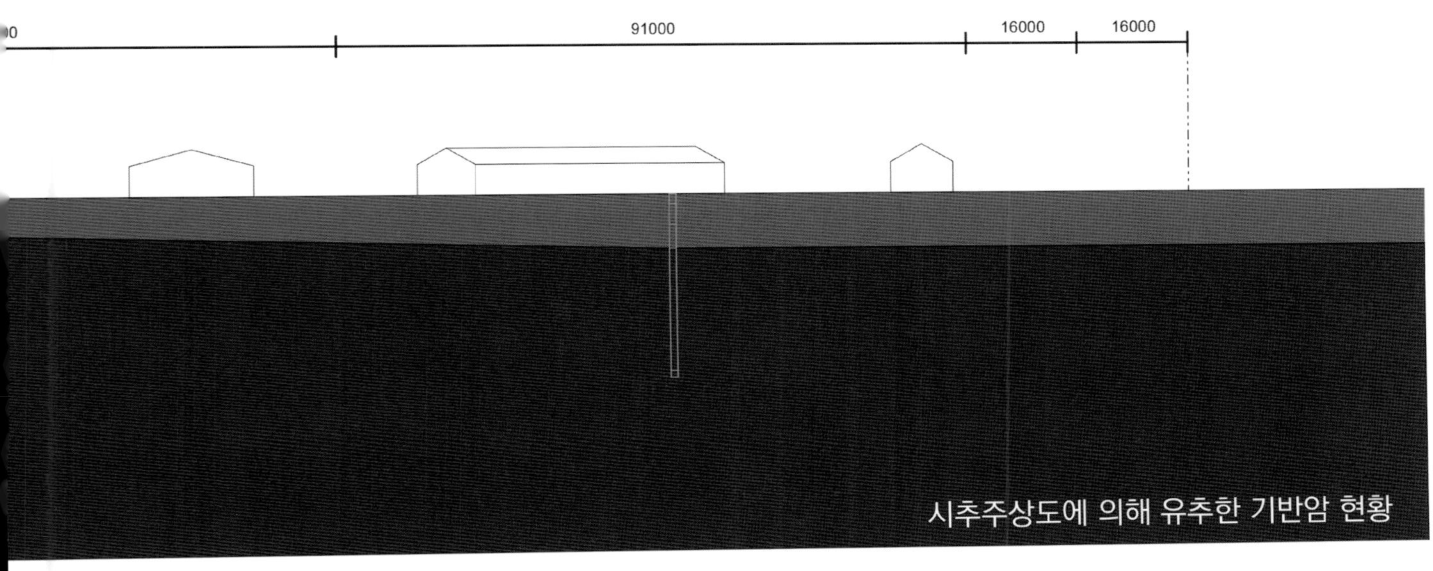

시추주상도에 의해 유추한 기반암 현황

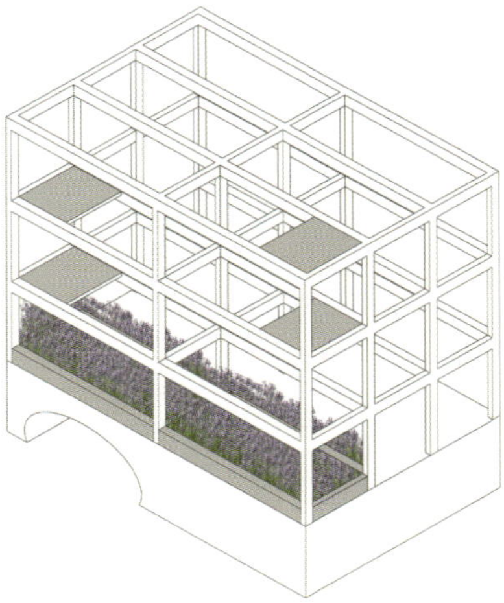

서식지 타입 A

물과 가까운 풀숲, 풀밭에 서식하는 새들을 위한 타입입니다. 잔디, 초화류만이 자랄 수 있는 생존최소토심과 생육최소토심을 고려한 화분으로, 새들은 마른 풀, 줄기 등을 쉽게 구해 둥지를 지을 수 있습니다.

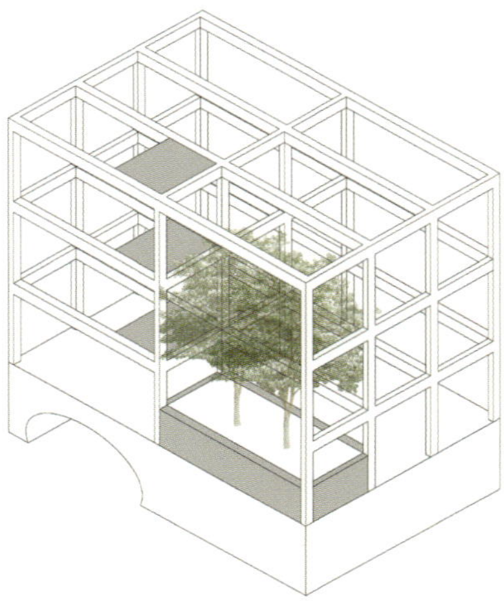

서식지 타입 B

나무 또는 지면으로부터 1m 높이에 서식하는 새들을 위한 타입입니다. 나무를 중심으로 한 양측면 플랫폼이 위치해 있어, 나뭇가지 등을 쉽게 구해 둥지를 지을 수 있습니다.

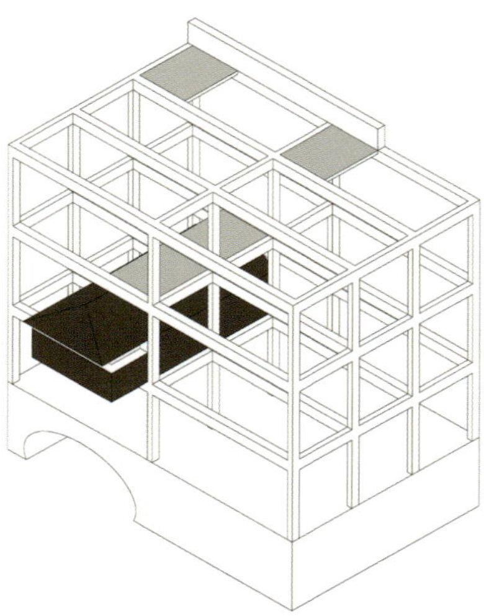

서식지 타입 C

건물 틈에 서식하거나, 스스로 둥지를 짓지 못해 다른 새의 새집에 서식하는 새들을 위한 타입입니다. 숨어서 새를 관찰할 수 있는 관측소를 기존 평화시장 구조에 삽입함으로 건물 틈을 제공합니다.

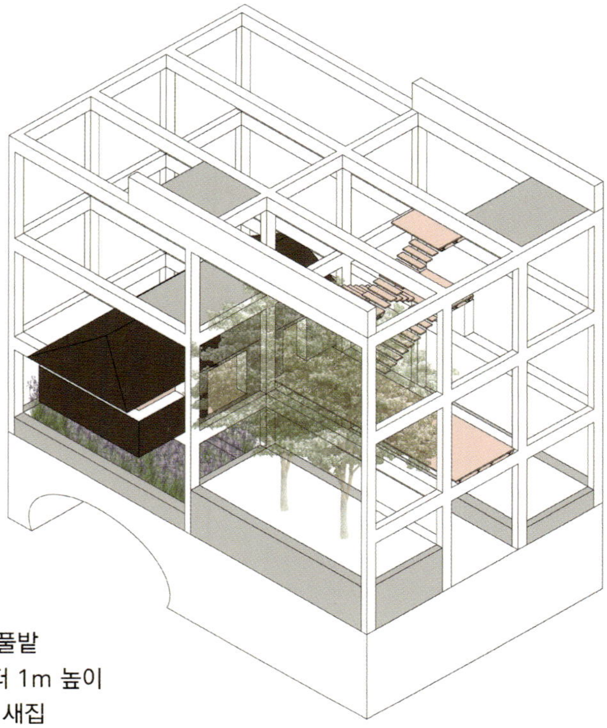

서식지 타입 종합

1. 물과 가까운 풀숲, 풀밭
2. 나무, 지면으로 부터 1m 높이
3. 건물 틈, 다른 새의 새집

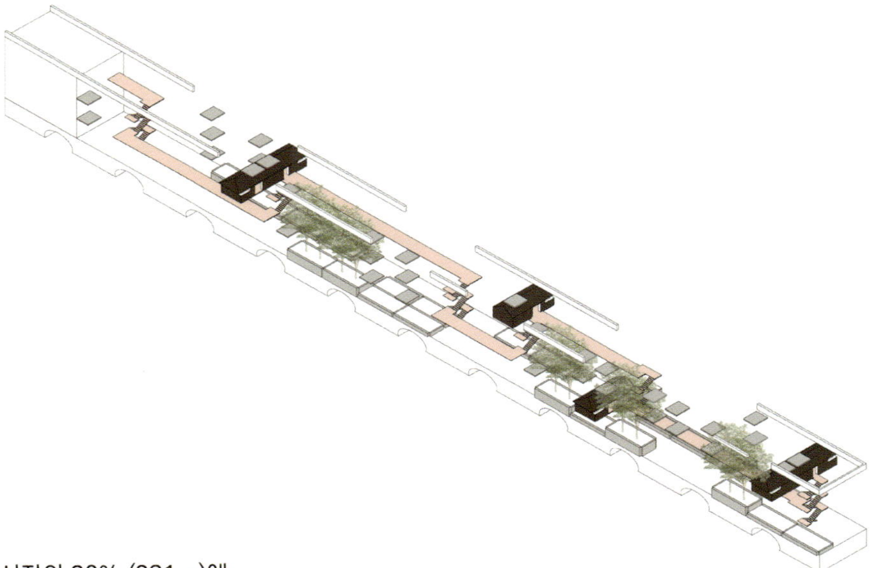

기존 평화시장의 30% (231m)에
구성한 보행동선, 새 관측소, 새 서식지

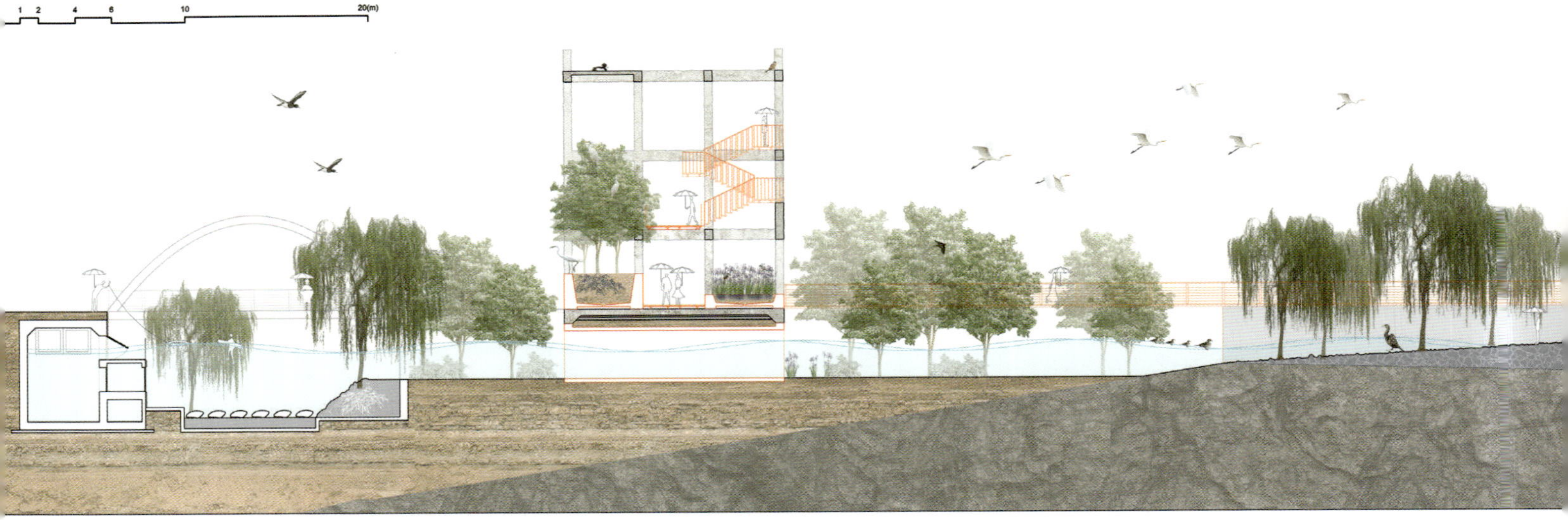

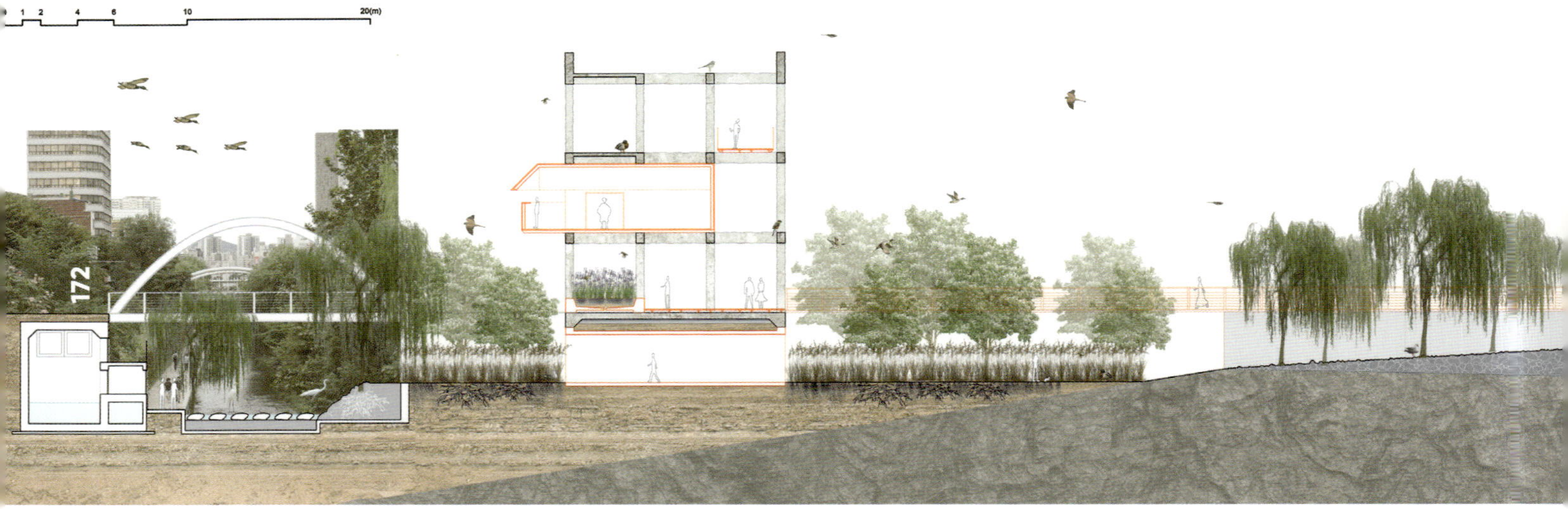

Project "Pyeongwha(peace) Market was blown up."

Based on historical incidents at Cheonggyecheon Stream, Namsan Mountain, Bamseom Island, and Jongmyo Street, it focuses on the conflicts between nature and the city that arose as attitudes toward nature changed in Seoul.
And I found that the background of Bamseom Island, which is considered a rewilding ecology, was manipulated through human intervention and manipulation.
"Among them, Bamseom Island is important in that it created "real" nature through human intervention.
Bamseom Island was an island of 1000 inhabitants, but it was blown up for the purpose of acquiring the stones needed for the development of Yeouido Island. Since then, continuous deposition has progressed in Bamseom Island, and has grown 10 times larger than the explosion, and it has become an ecologically important area. The background of this physical and ecological recovery of Bam Island is in the many dams that control the Han River. A series of dams have more accurately controlled the flow of the Han River than in the past, thus creating an environment where the flow velocity of the Han River will be slower than in the past and will naturally be deposited."
- In Jinho Kim's Research Presentation

Cheonggyecheon Stream also exists in a similar situation to Bamseom Island. In the past, Cheonggyecheon Stream was a natural river where surrounding tributaries gathered and flowed, but it was undergrounded for the construction of Cheonggye Overpass, which was rebuilt and now is an artificial stream that is supplied by a pump by mixing underground water and Han River water.
Nevertheless, Cheonggyecheon Stream is in the process of rewilding over time, with protected species inhabiting it.
In this background, I define "FREAKY, SIMULACRE ECOLOGY" as a new Seoul urbanism and propose the project name "Pyeongwha Market was blown up." at the U.S. Army Far East District Area.
Cheonggyecheon Stream has an important meaning from the perspective of birds. Birds need exposed cloths because they need to drink water or seek mud to make temporary shelters. As Seoul becomes urbanized, tributaries have become underground, except for Cheonggyecheon Stream. In this situation Pyeonghwa Market, which operation rate was 60 percent, partially exploded and returned part of site to migratory and resident birds in nearby mountains.
I conducted research on what birds already exist or are likely to visit in Cheonggyecheon Stream, where and how they live, what they need to build their nests, and how that material can get. And based on this research, I designed the human-bird coexistence type.
The deconstruction waste of Pyeongwha Market is used as a natural infrastructure, responds to disasters such as floods, and converts it into a fun, peaceful, and event.
Finally, Terra Incognita - unknown land, the area around the U.S. Far East District Area became open land for humans, birds, animals, plants, and even fish to occupy.

프로젝트 "평화시장은 폭파되었습니다."

청계천에서, 남산에서, 밤섬에서, 종묘거리에서 발생한 역사적 사건사고에 기반하여,
서울에서 자연에 대한 태도가 변화함에 따라 발생한 자연과 도시의 갈등을 조명합니다.
그리고 이 중 재자연화 중인 생태 (REWILDING ECOLOGY)로 여겨지는 밤섬의 배경에는
인간의 개입과 조작으로 (MANIPULATED) 형성되었다는 것을 발견하였습니다.
"이 중 밤섬은 인간의 개입을 통해 '진짜' 자연을 조성했다는 점에서 중요합니다.
밤섬은 1000명의 주민들이 살던 섬이었으나, 여의도 개발에 필요한 돌을 획득하기 위한 목적으로
폭파되었습니다. 이후 밤섬에 지속적인 퇴적이 진행됨에 따라, 현재는 폭파 이후 10배의 크기로
성장하며, 생태적으로도 중요한 지역이 되었습니다. 이러한 밤섬의 물리적, 생태적 회복의
배경에는 한강을 통제하는 수많은 댐과 보가 존재합니다.
일련의 댐은 과거보다 한강의 유량을 보다 정확하게 조절하게 되었고,
이에 한강의 유속이 과거보다 느려지며, 자연적으로 퇴적이 될 환경을 조성했습니다."
- 김진호 리서치 발표 中

청계천 또한 밤섬과 비슷한 상황 속에 존재합니다. 과거의 청계천은 주변 지류들이 모여 흐르던 자연하천이었으나, 청계고가도로 건설을 위해 지하화되었다, 다시 복개되며, 현재의 청계천은 지하수와 한강물을 섞어 펌프로 공급하는 인공하천입니다. 그럼에도 불구하고 청계천은 시간이 흐르며, 보호종이 서식하는 등 재자연화 과정 중에 있습니다.
이러한 배경 속 "FREAKY, SIMULACRE ECOLOGY"를 새로운 서울 어바니즘으로 정의하고, 미 극동 공병단 기지 일대 (U.S. Army Far East District Area)에서 프로젝트 "평화시장은 폭파되었습니다."를 제안합니다.
새(철새와 텃새)의 관점에서 청계천은 중요한 의미를 가집니다. 새에게는 외부에 노출된 천이 필요한데, 이는 새가 물을 마시거나 임시거처를 만들기 위한 진흙을 구해야 하기 때문입니다.
그러나 서울이 도시화되며, 청계천을 제외한 지천은 지하화 되었습니다.
이런 상황 속 60% 운영율을 보이는 평화시장을 일부 폭파시켜, 사이트의 일부를 철새들에게, 그리고 인근의 산의 텃새들에게 환원합니다.
청계천에 이미 존재하는 또는 방문할 가능성이 있는 새들은 무엇인지, 그들은 어디서, 어떻게 서식하는지, 둥지를 짓는다면 무엇을 가지고 둥지를 짓는지, 그 재료는 어떻게 획득할 수 있는지에 대한 리서치를 수행했습니다. 그리고 이를 바탕으로 인간-새 공존 타입을 설계하였습니다.
이때, 평화시장을 폭파하며 발생하는 건설폐기물을 자연적 인프라로서 활용하며, 홍수라는 재난에 대응하며, 재난임과 동시에 재미있고, 평화롭고 이벤트로 전환시킵니다.

마침내, 오랫동안 Terra Incognita - 미지의 땅의 상태로 존재했던 미 극동 공병단 기지 일대는 비단 사람 뿐만 아니라 새와 동물, 식물, 심지어 물고기가 점유할 수 있는 열린 땅이 되었습니다.

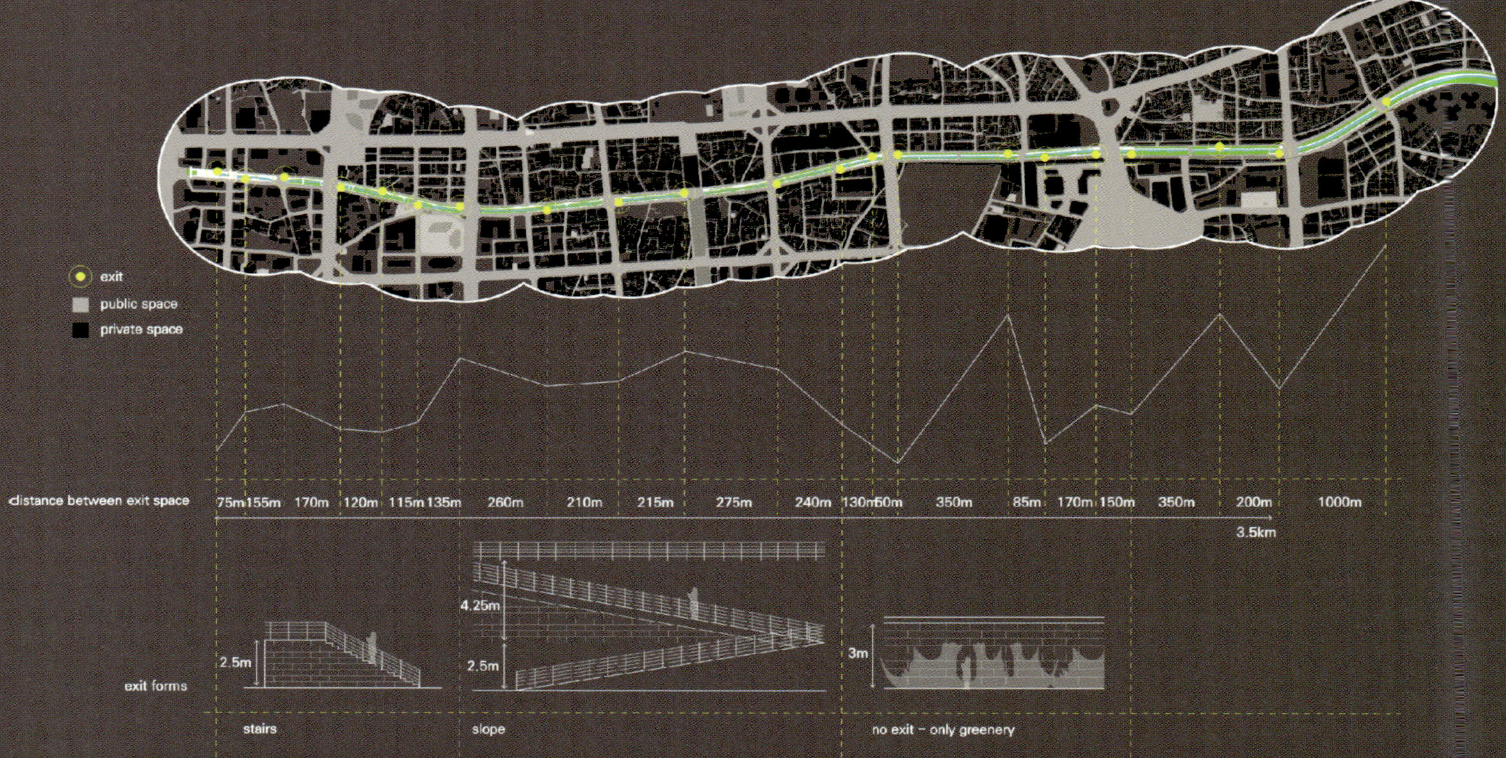

Site Analysis

Context study inside Jongro-gu and Jung-gu

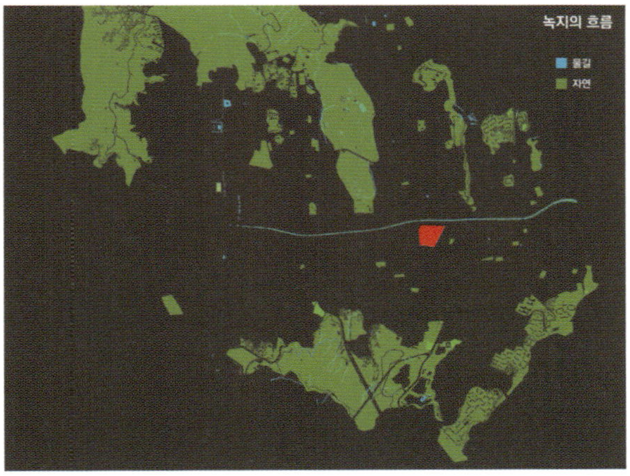

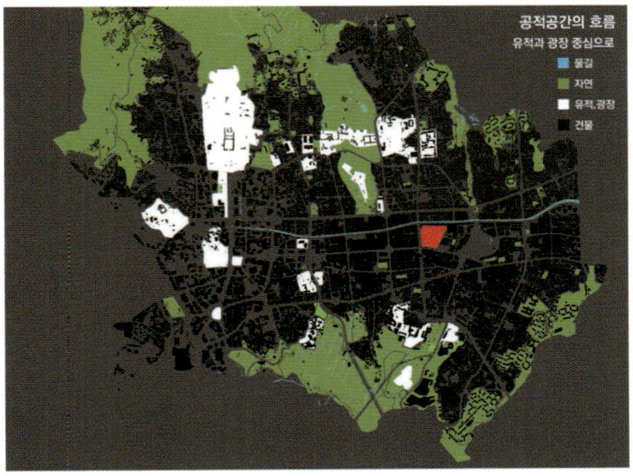

난전을 시전으로 - 그늘졌던 동대문 일대에 빛을 제공하다

DDP는 처음 지어질 때만 해도 반대가 극심했던 프로젝트였다.
당시에는 생소했던 자하 하디드라는 해외 건축가, 우리나라의 정서와는 어느 정도 거리를 지닌 UFO스러운 매스스트럭쳐 그리고 그 특유의 비정형적 구조로 인한 막대한 예산이 일반적인 국민들에겐 쉽게 이해되지 않는 것이었다.
DDP가 처음 지어질 당시의 반대여론은 약 70%에 달했다.
많은 이들이 설계와 더불어 한 가지의 문제를 꼽았다. 그것은 공간이 지어지는 목적의 불명확성이었다.

그로부터 약 10년이 지난 지금 DDP가 실패작으로 끝났다고 말할 사람은 없을 것이다. 확실히 DDP는 누구나 알 정도의 랜드마크로써 동대문 주변 디자인 플라자를 정의하는 공간이 되었다. 의류와 예술을 중심으로 많은 이벤트들이 대표적으로 패션쇼 등이 - DDP에서 개최되며 대로가 살아나고 주변의 쇼핑몰이 고개를 들었다. 구글맵의 검색 추이를 보면 DDP가 개방된 이후로 동대문과 관련한 검색량이 급증하였음을 볼 수 있다. 이전까지 동대문의 이미지는 좋든 나쁘든 소상공인들과의 끝없는 흥정을 통해 출처가 불분명한 물건을 (자칫하다간 비싼 값으로) 구매하는 시장이었다. 값싼 옷을 구매하기 위해 가는 곳이었지만 손해를 볼 확률도 그만큼 높은 곳이었다. DDP를 통해 대로 주변 일대는 패션의 메카가 되었고, 그 스케일에 준하는 두타 등의 대형 쇼핑몰이 기존 시장의 역할을 대신하여 관광객을 불러모았다. 동대문 디자인 플라자라 불리는 영역은 새로운 세대를 위한 공신력을 갖춘 광장으로써 그만의 가치를 확보했다.

여기까지가 평소 동대문 일대와 관련이 없는 삶을 살아가는 제 3자의 관점이다.
조금 더 깊숙이 들어가보면 동대문의 이러한 변화가 실제 지역의 생태에 얼마나 직접적으로 도움이 되었는지 관련하여 의문이 생긴다.
근대, 심지어 조선 시대까지 거슬러 올라가보아도 동대문시장은 자연적으로 형성된 군집의 시장이었다. 양반과 귀족들이 구매할 수 있었던 시전이 아니었다.
한양의 외곽에 거주하던 병사와 일반 상민들이 정부의 허락 없이 눌러앉아 생필품을 제작하고 판매하던 난전이었다. 일제 강점기 시절에는 전차로와 기존의 청계천을 허물고 지은 고속도로라는 교통축이 형성되었다.
이런 요소들이 부지의 군집성을 강화했다. 서민들은 자연스럽게 교통로의 주변에 거주하며 시장을 형성했다. 특히 1950년대 초반의 6.25 전쟁 전후에는 갈 곳을 잃은 난민들이 교통로 주변에 하나둘씩 모여들며 판자촌을 꾸렸다.
'평화'시장, '통일'상가라는 시장의 이름들은 후방으로 밀려나 평화로운 삶을 꿈꾸었던 소상공인들의 마음을 보여준다. 동대문의 키워드는 허가 받을 여유조차 없다는 '불법성'과 그 안에 담긴 살아남으려는 마음이었다.

성장세에 제동을 건 것은 동대문 일대를 받치던 축들을 퇴보 시킨 2000년대 이후 두 차례의 사건이었다. 첫 단계는 청계천 복개 공사였다.
청계천 복개는 근대 이후로 메워진 도로를 허물고 원래의 청계천 개울을 서울에 돌려놓고자 하는 정책이었다. 청계천 복개공사는 처음부터 잡음이 많았다.
도로를 허무는 과정에서 수없이 많은 상인들이 삶의 터전을 잃고 도로에서 멀리 떨어진 후방으로 밀려났다. 제대로 된 신고를 거치지 않고 불법적으로 사업을 꾸리던 이들이었기 때문에 그들은 후속 대책을 받지 못한 채 사라졌다.
더 큰 문제는 청계천 복개공사가 의류시장이 평화시장 인근에 발달했던 주요 원인을 없애 버렸다는 것이었다. 동대문이 큰 시장으로 성장할 수 있었던 것은 그곳이 창신동에서 동대문시장까지 이어지는 의류 생산의 종착점과 도로를 지나는 일반인들의 동선이 만나는 교점이었기 때문이었다.
청계천 공사는 기존의 도로 가운데에 폭 8m, 높이 4m의 선큰 통로를 만들어 일반인들이 지나가도록 했다. 기존의 지면 레벨은 물류를 운반하는 트럭들이 좁아진 차로를 힘들게 오가는 형태로 바뀌었다.
기존의 소비자들은 힘들게 걸어 평화시장을 찾기보다는 지하철역을 거쳐 바로 DDP 등의 파사드 지역으로 이동했다. 고의적으로 낮춘 교통의 레벨은 그것이 중심이 되었던 도시의 전체 시스템도 그만큼 낮추었다.

Lighted Areas: Pedestrian-accessible district along the Cheonggaechun Area

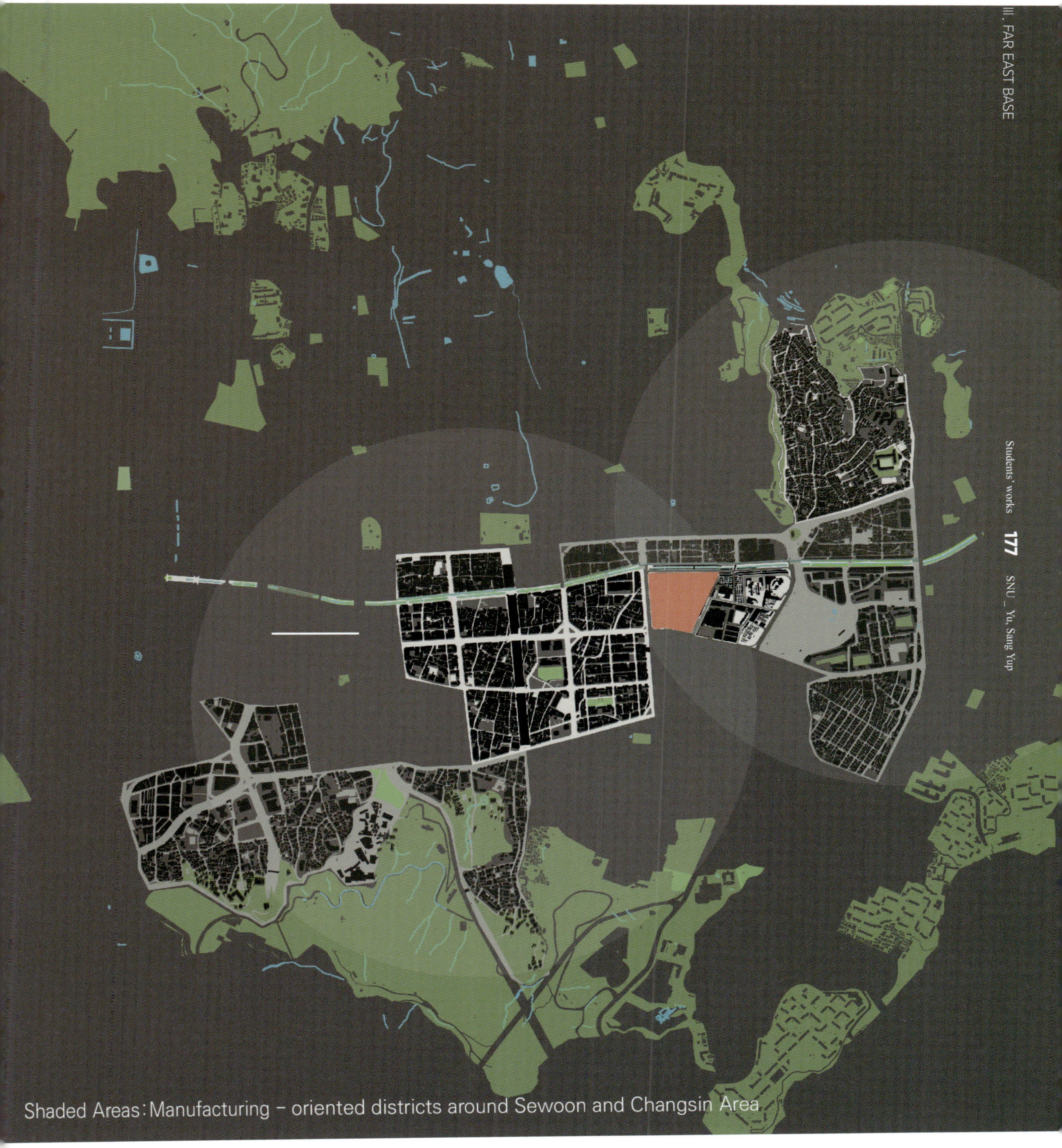

Shaded Areas : Manufacturing – oriented districts around Sewoon and Changsin Area

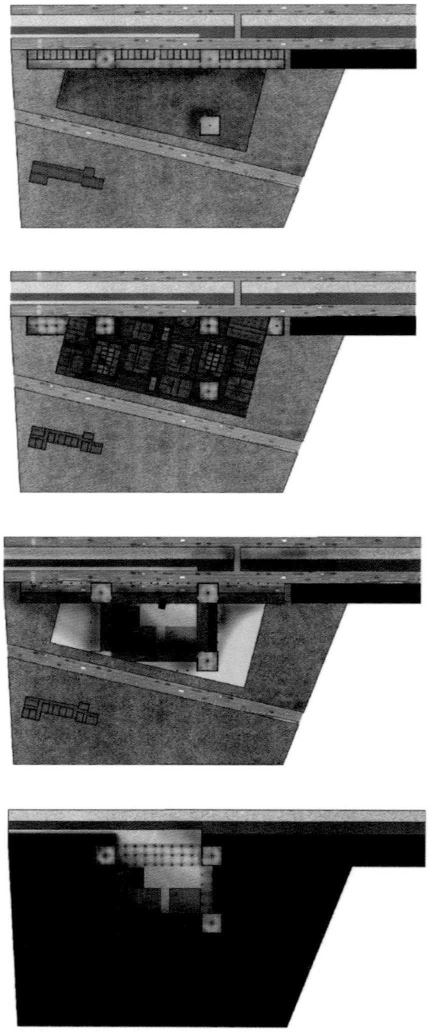

first floor below ground level

1 entrance
2 shaded garden
3 slope
4 stream

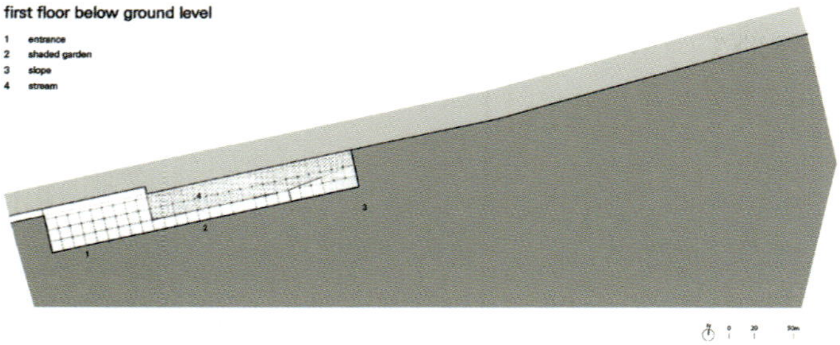

first floor

1 cafeteria
2 kindergarten
3 greenery

5 cooperation canopy
6 open lecture hall
7 office

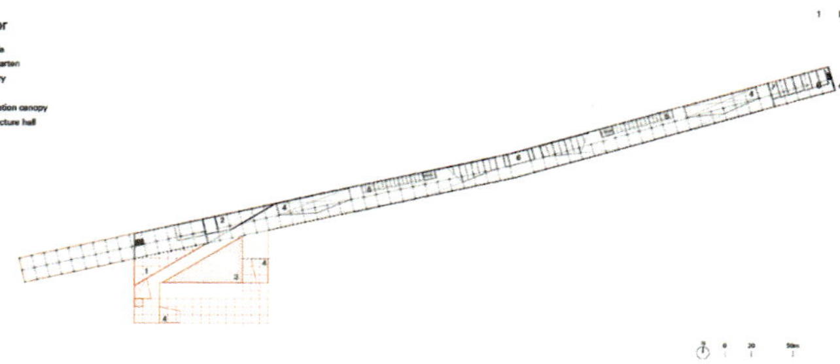

ground floor

1 car road
2 parking lot
3 loading dock
4 distribution center
5 steel manufacturing studios
6 open studios
7 slope
8 lecture hall
9 slope
10 commercial canopy
11 visitor plaza
12 information center
13 slope

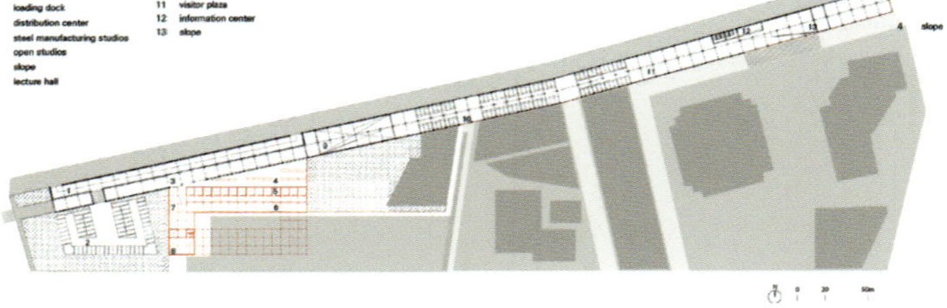

second floor

2 timber manufacturing studios
3 lecture hall
4 sewing studios
5 cutting studios

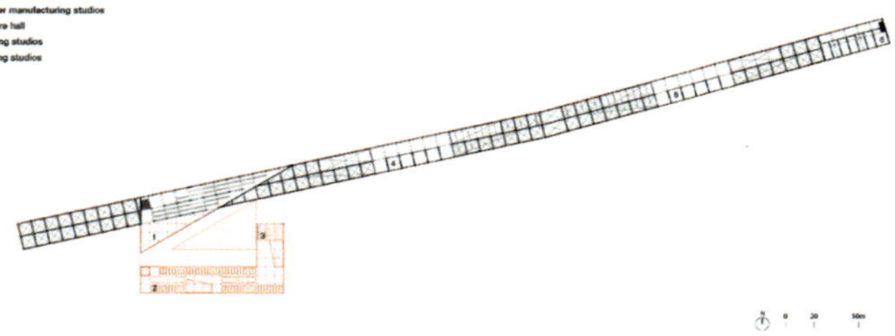

DDP가 지어지면서 바뀐 소비 시스템도 문제였다. 정확히 따지자면, DDP가 상징하는 새로운 소비 시스템에 정작 동대문의 생산시스템은 포함되어 있지 않았다는 것이 문제였다. 구글 검색 추이를 보면 DDP를 방문하는 방문객의 대략적인 경로를 볼 수 있다.
그들은 '특정 이벤트'를 보러 DDP를 방문했다가 '그에 만족하는 물건'을 사러 주변의 '두타' 혹은 '백화점'과 같은 대규모 상업공간으로 이동한다. 대규모 소매상점들은 대량으로 물건을 떼어온다. 물건의 95%는 외부 공장에서 들여온 것이다. 그 중 50%는 한국 내부 대구와 부산 등의 지방, 45%는 중국과 인도 등의 개발도상국이다. 인접한 동대문의 생산단지에서 만들어진 물건은 어디에도 포함되어 있지 않다.

무작정 동대문의 물건을 떼어오라고 강요할 수는 없다. 쇼핑몰의 사업 방식은 소매시장의 그것과 사뭇 다르기 때문이다.
쇼핑몰에 입점한 브랜드는 대부분 일정 규모 이상의 브랜드로 마진을 남기기 위해 할당된 판매량을 만족해야 한다. 동일한 제품이 기계화된 공정을 통해 대량으로 만들어져 보급될 수 있는 공장이 필수적이다. 반면 창신동에서 제작된 의류를 비롯한 동대문시장의 상품들은 의류의 경우 9-10명, 기계설비의 경우 3-4명의 적은 인원이 모여 개별적인 과정을 통해 생산되는 과정을 거쳐 만들어진 상품들이다. 각 상품이 모두 다르다. 불량률도 높다. 전략상 필수적인 통일성과, 품질에 대한 보장이 무너진다.

쇼핑몰에서 동대문 인근 생산단지의 물품들을 들여올 수 없는 또다른 이유는 자격요건의 부족에 있다. 창신동에 위치한 의류 생산 공장의 70% 이상은 필수 자격증을 따지 못한 경우로 물건을 생산함에 있어 필요한 자격을 충족하지 못한다. 이들은 국가적으로 생산시설로 등록되어 있지 않다. 우리가 알고 있는 사실과는 다르게 지도상으로 주거시설 및 근린시설로만 등록되어 있을 뿐이다. 그 이유를 알기 위해선 다시 거슬러 올라가 동대문 인근이 제조시설로 부상했을 시기를 보아야 한다. 우리나라가 제조업 강대국이었던 7-80년대, 제조업자로 일할 수 있는 필요조건은 따로 존재하지 않았다. 이때 동대문으로 몰려든 노동자들이 지금 동대문 인근에서 제조업자로 일하고 있는 노동자의 80% 이상으로 4-50대 사이의 나이대를 가진 고졸 이하 학력의 노동자들이다. 시설과 기술 수준이 21세기의 그것을 만족시키지 못하므로 그들은 동대문시장 너머로 진출하지 못한다. 그들이 상대적으로 넓은 면적을 차지하고 있음에도 불구하고 그에 준하는 사회적 영향력을 갖지 못하는 이유다.

그렇다고 언제까지 이들을 좌시할 수 없는 이유는 이들의 퇴보가 지역 자체의 슬럼화로 이어지기 때문이다.
타 지역과 달리 동대문 지역 거주자의 나이 편향은 과도하게 한쪽으로 치우쳐져 있다. 지역의 대표적인 거주자들이 임하는 분야가 지역 경제 체제에 기여하지 못한다는 사실은 지역의 수명이 얼마 남지 않았다는 것을 의미한다. 실제로 평화시장 인근의 구역은 중앙의료원과 주민회관 등의 편의시설이 위치해 있음에도 불구하고 슬럼화가 진행되어 인근의 두타와 비교했을 때 확연한 유동인구의 차이를 보인다. 중앙의료원의 환자들 몇 명만이 대낮에도 서성이며 담배를 필 뿐이다. 지역을 부활시키기 위해선 동대문의 대부분 지역을 차지하는 중심 시스템을 없애는 것이 아닌 유지하면서 새로운 세대의 인구를 불러들여야 한다. 그러기 위해선 그들의 시스템에 21세기에 맞춘 새로운 자격요건을 부여하는 것이 필수적이다.

평화시장 뒤편에 커뮤니티 칼리지가 세워질 이유는 여기에 있다. 언뜻 보면 이곳은 창신동 쪽 의류 생산 공정의 최종점 너머의 지점이자, 세운상가 쪽 기계생산 공정 바깥의 공간으로 두 공정과 모두 유리된 구역처럼 보인다. 하지만 이들에 직접적으로 참여하지는 않으면서 인접해 있다는 특징은 결국 이 모두를 엿볼 수도, 진입할 수도 있음을 의미한다. 본 구역은 기존 동대문 시장 생산 공정의 파사드로 여겨졌던 평화시장의 이면에 가려진 그늘이 아닌, 새로운 파사드로써 작동할 것이다. 평화시장의 주민들이 생산공정에 필요한 자격요건을 갖추는 학업의 공간으로써 동대문 일대를 쇼핑몰에 준하는 공신력을 갖춘 공간으로 탈바꿈시킬 것이다. 또한 젊은 세대가 독립적으로 원하는 제품을 수주하고 협력을 통해 생산에 간접적으로 체험할 수 있는 공간으로 만들어 새로운 혈기를 수혈시킬 것이다.

B'
B
0
5
10m
section B - B'

A
A'

개미굴에 학교를 짓는다는 것

앞서 우리는 커뮤니티 칼리지가 동대문 지역에 왜 필요한지를 알아보았다. 하지만 더 명확한 설계를 위해선 본 지역에 학교라는 시설이 어떠한 의미를 갖는지를 정의해야 한다. 다른 지역과 비교하여 동대문 지역은 독특한 모습을 갖는다. 제조업에 뿌리를 둔 특성상, 동대문은 적은 인원이 모인 사업체가 독립적으로 기능하는 군집의 형태를 띤다. 다른 지역의 시장경제에 비하여 경쟁이 비교적 적게 일어난다.
일례로 미대생이나 건축학도가 을지로의 제조업자에 찾아가면 전문분야가 아닌 경우 근처의 다른 사람을 소개받는 경우가 많다. 이는 전형적인 시장경제체제로선 이해하기 힘든 부분이다. 을지로와 창신동의 제조 시스템이 네트워크로 짜여져 하나의 공장 아래 부품으로써 기능하기에 가능한 일이다. 동대문은 자본주의 국가의 산물임과 동시에 그 최전성기였던 1970-80년대 후반 근현대의 집단주의가 공간적으로 구현된 지역이다. 하나의 부품이 작동하기 위해선 다른 부품이 있어야 한다.

제작하는 분야상, 창신동은 선형적이고 을지로는 비선형적인 배치를 보인다. 효율적인 교육기관을 설계한다면 제작방식을 따라 자연스럽게 생겨난 이러한 모습을 그대로 흉내내어 공간으로 만드는 방법이 있을 것이다. 하지만 이러한 접근은 대학교라는 시설의 중요한 측면을 간과한 접근이다. 동대문 지역을 반영함에 있어 중요한 개념은 배치상의 차이보다도 제작 종사자 모두가 고정된 위치에서 24시간 기능할 수밖에 없다는 사실이다. 제조산업 전체가 컨베이어 벨트와 같이 작동해야 한다. 주어진 작업만을 성실히 수행하는 것은 주민들에게 평생에 걸친 과제였다. 하지만 각각의 제조 분야에 맞추어 가장 효율적인 방식으로 진화했다는 것은 그만큼 창의적인 작업이 자라날 순 없다는 한계가 됐다. 21세기 들어 동대문이 변화하는 사회에 유연하게 대처하지 못하며 퇴보한 이유다.

개념적으로 대학교는 그 대척점에 위치한다. 대학시설의 가장 큰 정의는 '움직임'이다. 움직임이 초중고로부터 대학교를 구분짓는다. 학생들은 항상 앉아 있다. 움직이는 것은 선생님이다. 학년이 진급할수록 컨베이어 벨트처럼 고정지가 이동될 뿐, 그동안은 고정된 장소에서 정해진 작업을 수행하는 것이 기본이다. 반면 대학교는 수업 분야별로 강의동이 고정되어 있다. 학생들이 수업에 맞추어 이동한다.
이런 이동의 과정 속에서 자신의 분야가 아닌 타과생을 만나기도 하고, 일반적인 수업만으로는 교류할 수 없는 교수님과도 대화하며, 타인과 작업, 그리고 관심을 공유하게 된다. 수업만큼이나 중요한 것은 이런 자유로운 움직임 속 더 넓은 세상과의 교류를 통해 자신의 세계를 만드는 일이다. 수업을 통해 기술과 지식이 증진되고, 움직임을 통해 창의력이 증진된다.

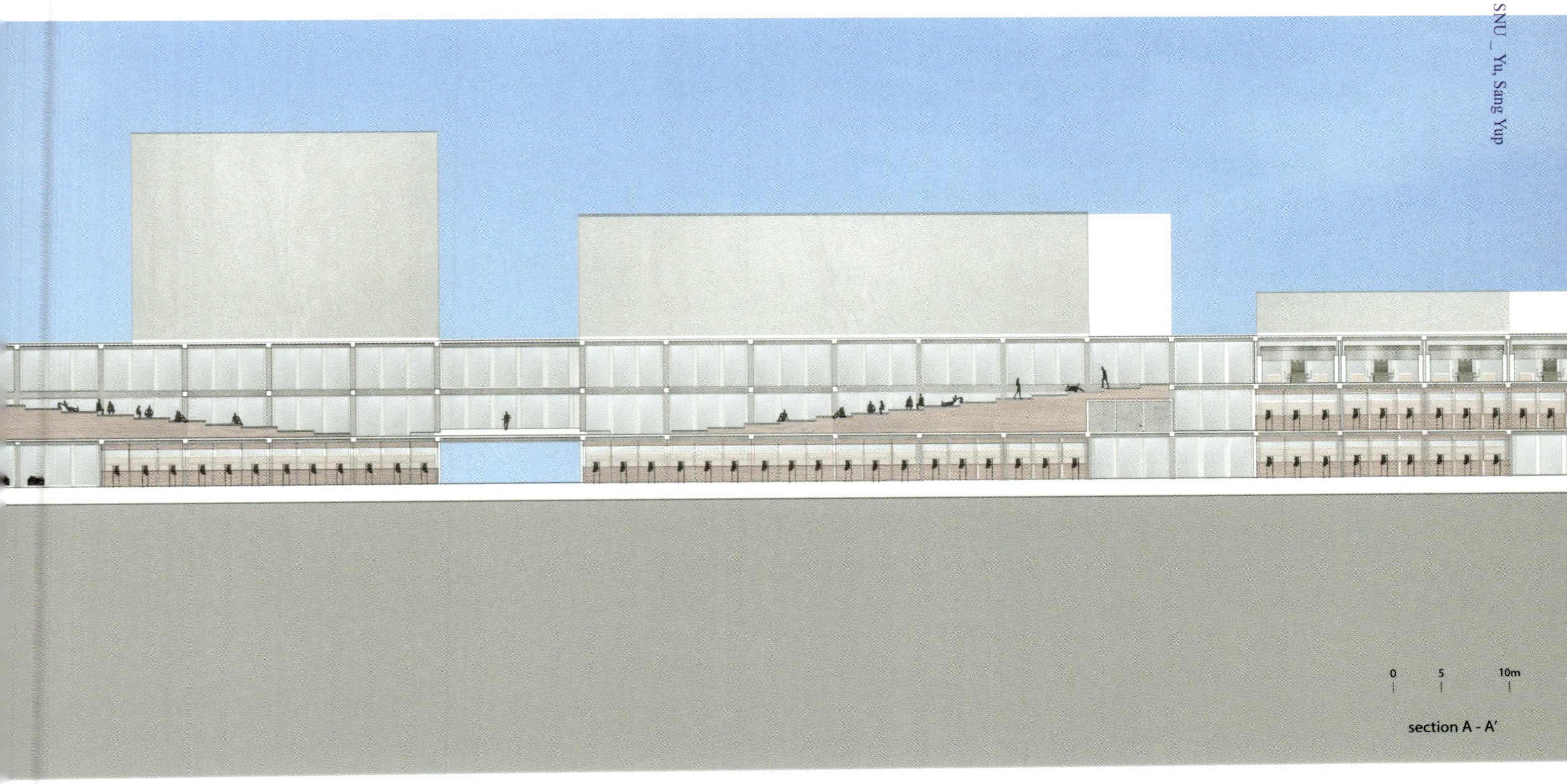

section A - A'

움직임은 선적이고, 시선은 공적이다. 우리는 공간으로 환경을 인식하지만 선으로써 움직일 수밖에 없다. 결국 자유로운 움직임과 교류가 전반을 차지하는 일반적인 대학시설의 구성(임의적으로 배치된 강의홀과 그 사이의 넓은 공적 공간들)을 산업적인 환경 속에 위치한 커뮤니티 칼리지에 그대로 반영할 순 없다. 동대문 지역의 네트워크는 효율성을 바탕으로 자라난 것이며 하나의 전체적인 그림을 구성한다. 그 중앙에 이러한 네트워크를 무시하는 임의적 공간을 넣게 된다면 전체가 깨질 것이다. 대학의 전체 배치는 결국 사이트의 논리를 따라 선적인 움직임을 이어 나갈 것이다.

대신 익숙한 구성 속에 새로운 요소를 충돌시켜 가능성을 만든다. 커뮤니티 칼리지로써 학교의 잠재력을 극대화시킬 수 있는 방법은 결국 고정된 움직임을 무리하여 깨트리기보다는 움직임을 넘어서는 공적인 시야를 극대화하는 것이다. 기존의 주민들이 보지 못했던 부품 너머 사람을 목격할 수 있도록 시야가 컨테이너 벨트 안으로 관입된다. 움직임은 최대한 선적으로 통일하되 시선이 넓어진다.
이로써 익숙하되 새로운 경험을 만들어주는 공간이 생겨난다.

명확한 네트워크로 짜여진 스튜디오와 세부 강의실과 달리, 식당, 대강당, 도서관, 회의실, 상점은 넓은 외부 보이드에 흩뿌려져 있다는 것이 특징이다. 이 보이드는 어떤 부분은 부지 안에, 어떤 부분은 부지 밖에 걸쳐져 물리적인 입장이 가능한 곳과 불가능한 곳이 시각적으로 불분명하다. 마치 넓은 시야에 프로그램들이 임의로 떨어져 있는 것처럼 보인다. 이러한 구성은 공간을 실제보다 크게 보임과 동시에 서로 다른 프로그램들이 선적으로 배치되어 있음에도 불구하고 하나의 공간 안에 있는 것처럼 보이게 한다.
선적인 스튜디오 출구에서 나와 하나의 거대한 공에 있는 셈이다.

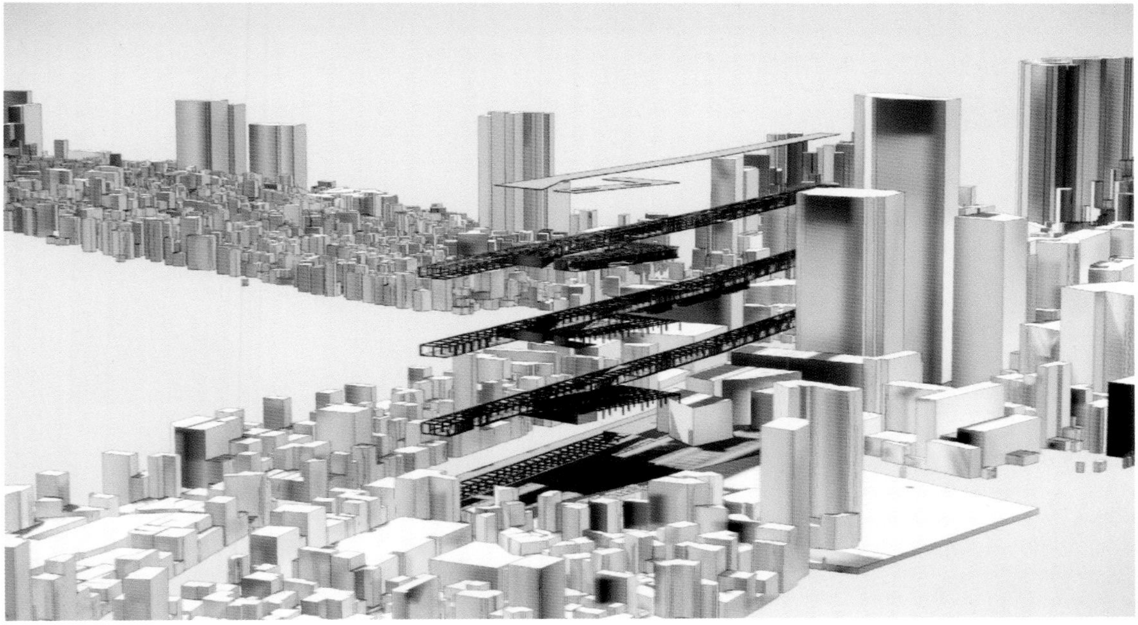

어려운 것은 기존의 평화시장 부지에 위치한 의류동이다. 기존의 부지가 워낙 좁고 긴 형태이기 때문에 스튜디오만 채워넣어도 건물이 꽉 찬다. 보이드가 관입될 여지가 전혀 없는 건물이고 너머의 시설들과 제조동이 전혀 보이지 않는다. 이로 인해 스튜디오 옆에 통로를 설치하게 된다면 통로 하나만으로 꽉 차서 결국 통로가 보이드가 아닌 통로로서의 목적을 지닌 공간으로밖에 남지 않게 된다. 따라서 의류동은 가로상의 접근보다는 수직적인 접근을 취해야 한다. 결국 보이드는 수직적인 뚫림에서 수평적인 뚫림으로 바뀌는 분기점이 생길 것이다.

의류동의 뚫림이 제조동까지 보이게 하려면 스튜디오는 3층, 진입로는 2층 전체를 뻥 뚫어놓아서 전체를 비워야 한다. 그리고 통일상가와 동화상가의 2층을 투명하게 만들고 상점과 학교를 이어붙여 공에 포함시킨다. 이 때 조금 떨어져 있는듯한 모습을 컨셉으로 삼는다.

공간(스튜디오, 강의실, 화장실)의 디자인 컨셉은 컨베이어 벨트로 서로 인접해 있고 휴먼 스케일인 반면, 보이드(강당, 식당, 통로, 도서관, 상점)의 디자인 컨셉은 창의적 우주로 서로 떨어져 있고 스케일이 더 크다. 공간의 세 동은 컨베이어 벨트처럼 구성을 취하지만 동시에 보이드를 기점으로 떨어져 있다. 하나의 동에서 다른 동으로 가기 위해선 보이드에 발을 담가야 한다.

제조동의 경우도 동일하다. 제조동의 기계제조 스튜디오들은 3층에 위치한다(공간들은 모두 3층과 1층에 있다). 2층은 보이드로 전체를 비워놓았다. 2층의 대형강의동도 결국 보이드에 있기 때문에 통행을 방해하지 않는 선에서 별도의 벽 없이 기둥만으로 구분된다. 따라서 2층의 평면은 마치 아무런 구회도 되어있지 않은것처럼 보이게 될 것이다.

REPROGRAMMABLE CACES

From small scale to large scale

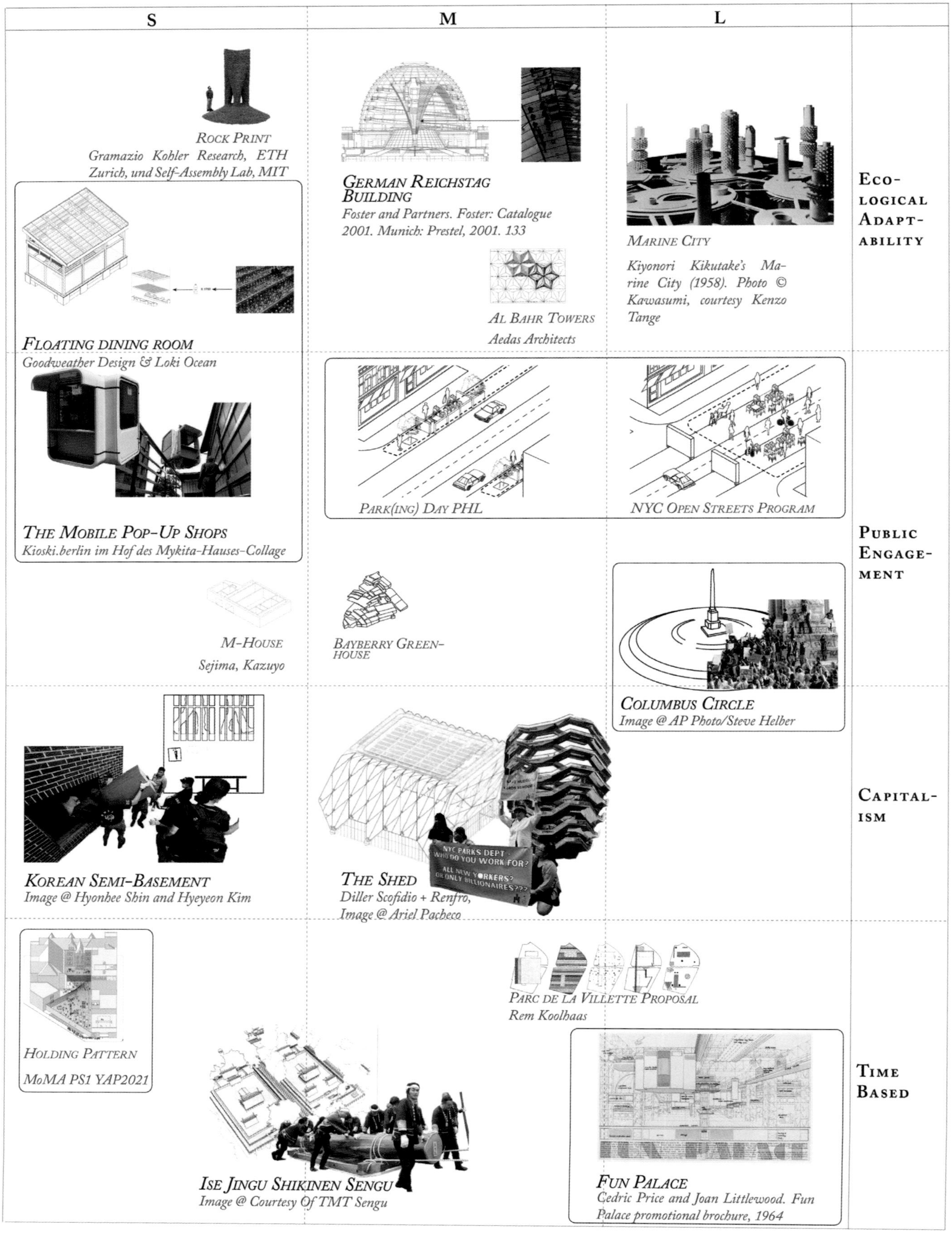

USERS ACTIVITIES ANALYSIS

BASED ON DONGDAEMUN GARMENT INDUSTRY

SIGNIFICANT BUILDINGS
- *A. food market*
- *B. hotel*
- *C. park*
- *D. hospital*
- *E. tourist center*
- *F. dongdamen market*
- *G. landmark*
- *H. school*

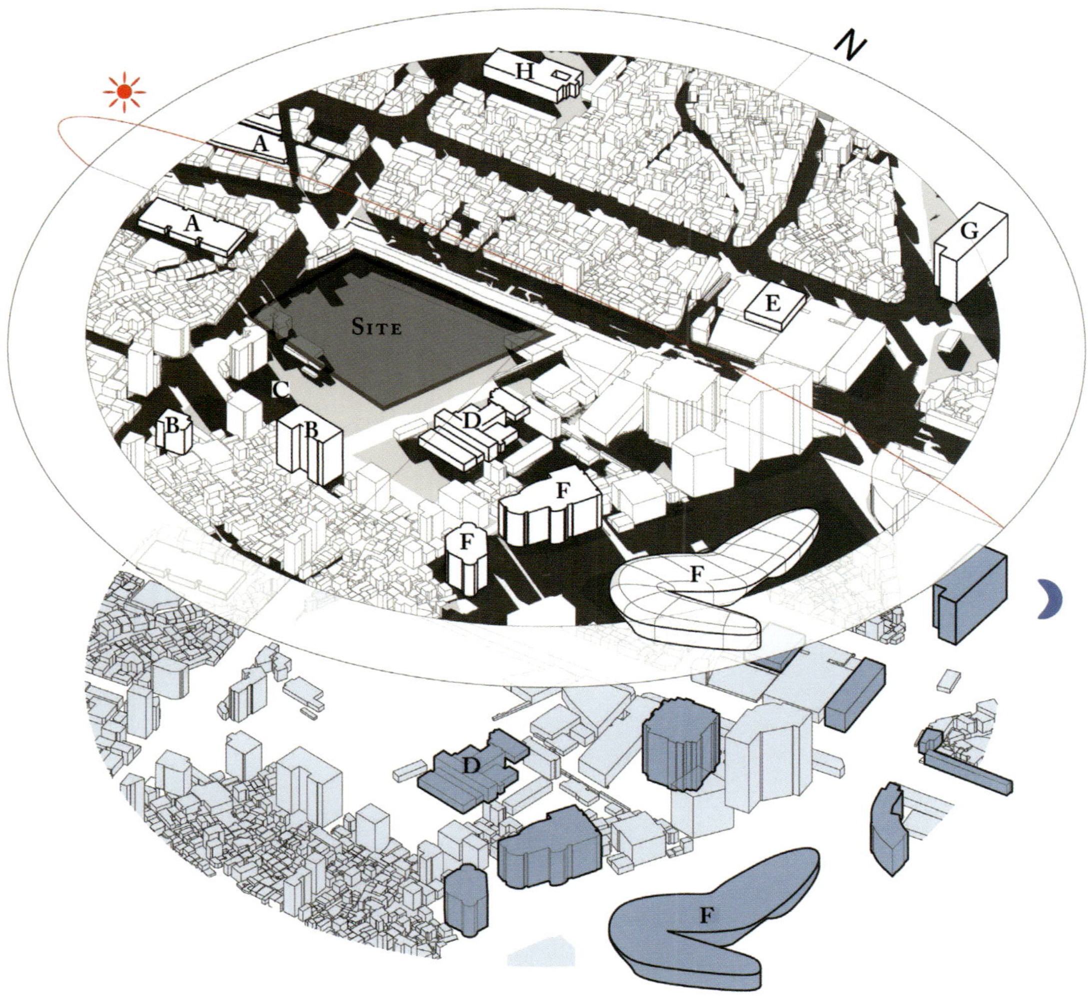

USERS
day:
- *food place owner*
- *tourist*

night:
- *designer*
- *freelancer*
- *shop owner*
- *logistics worker*
- *garment worker*

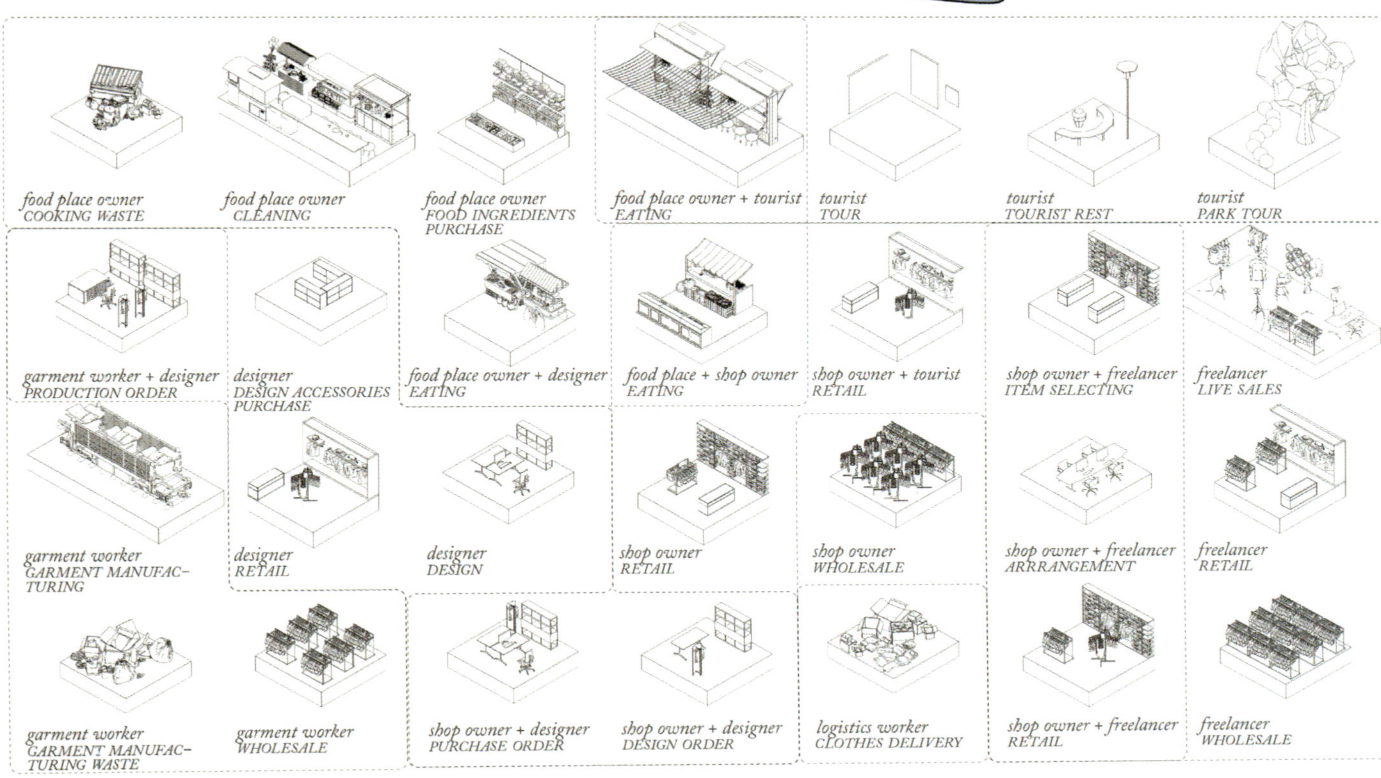

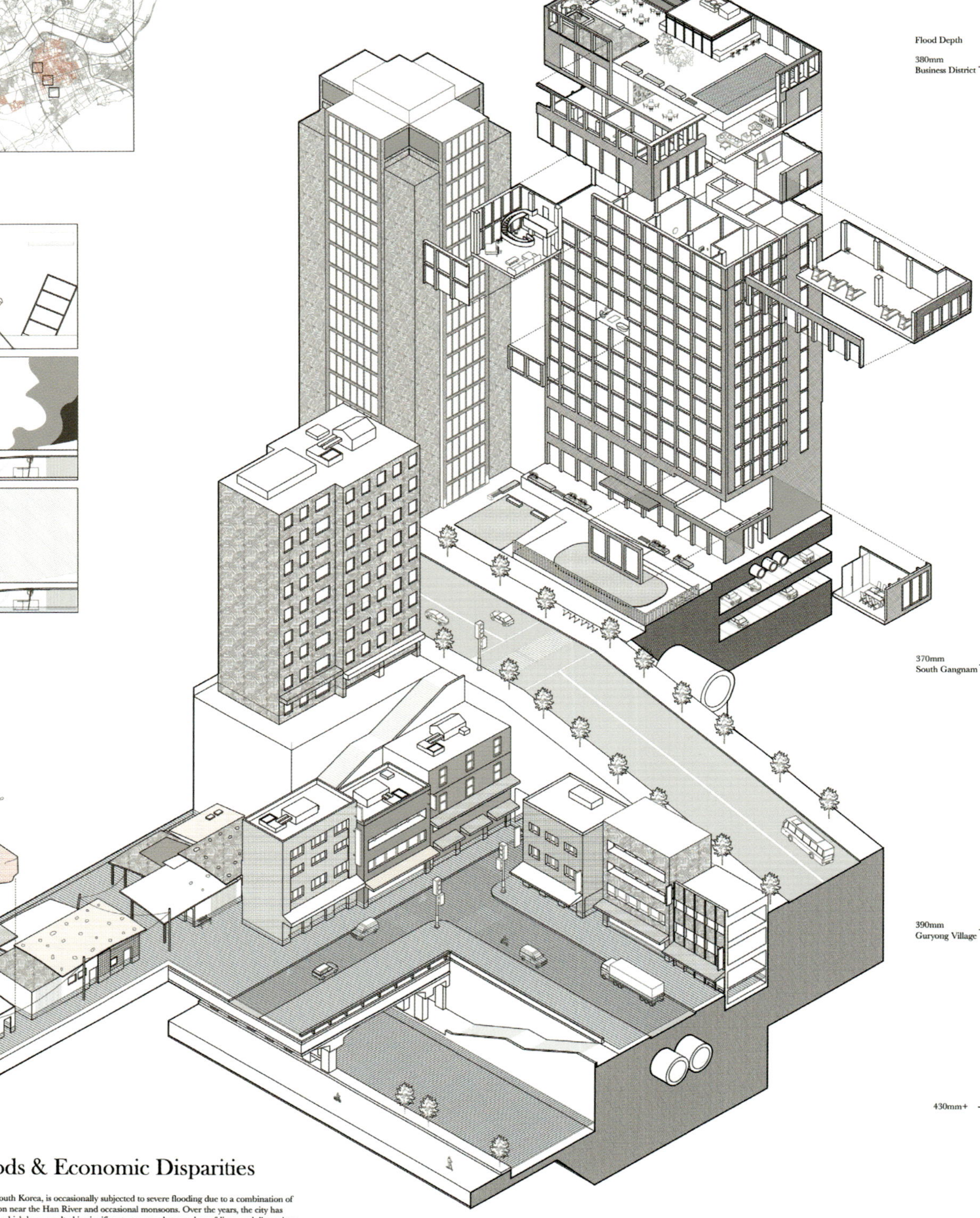

Seoul Floods & Economic Disparities

Seoul, the capital city of South Korea, is occasionally subjected to severe flooding due to a combination of factors, including its location near the Han River and occasional monsoons. Over the years, the city has experienced several floods, which have resulted in significant property damage, loss of lives, and disruptions to daily life.

Economic inequality is a concern not just in Seoul but worldwide. In Seoul, the rapid development and modernization over the past few decades have brought about substantial economic growth. However, this growth has not been evenly distributed, leading to disparities in wealth and opportunities among its residents. The Seoul Metropolitan Government has made efforts to address economic inequality, such as by improving public housing, offering job training programs, and providing social services. However, achieving a more equitable distribution of wealth and opportunities remains an ongoing challenge.

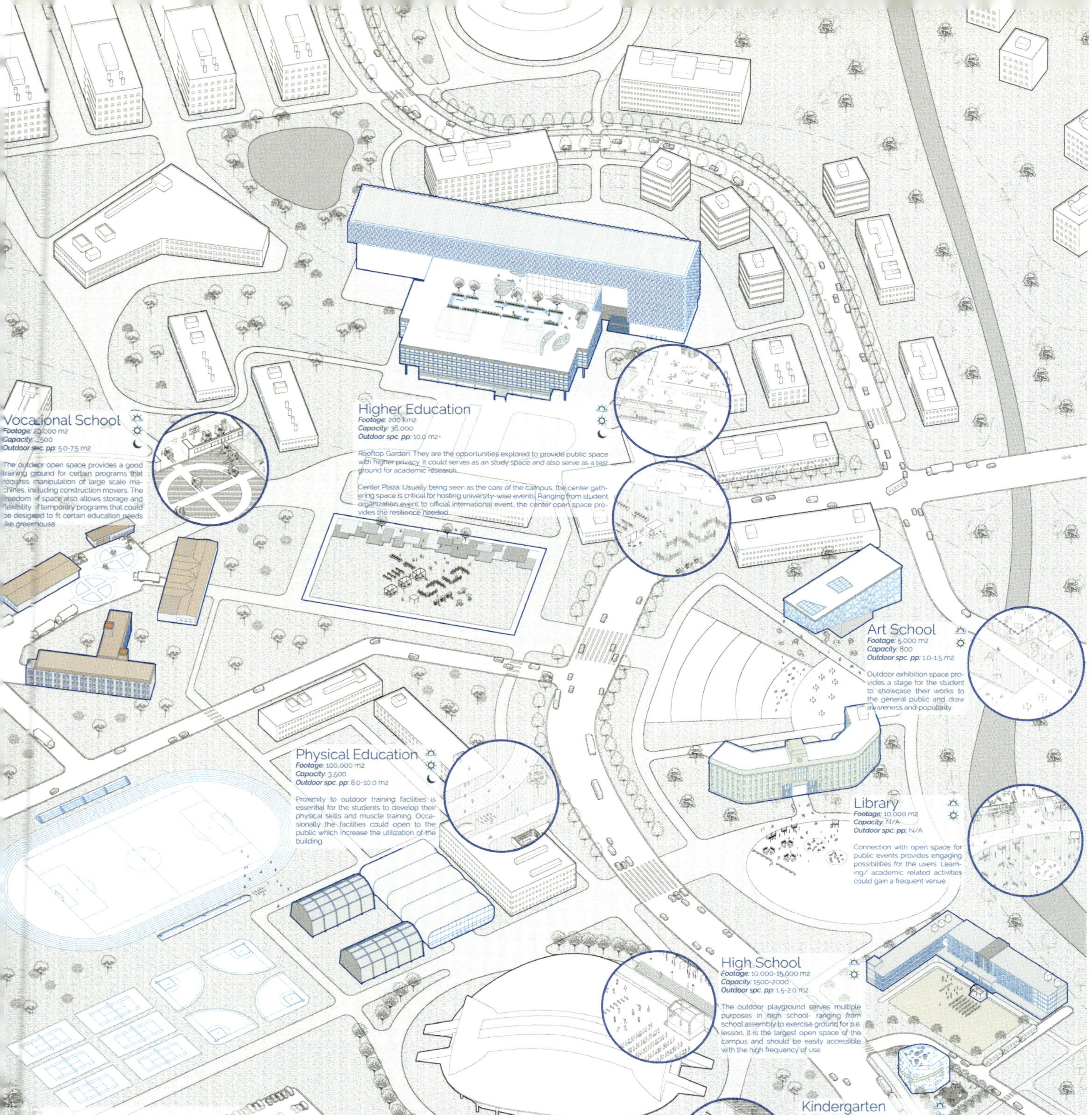

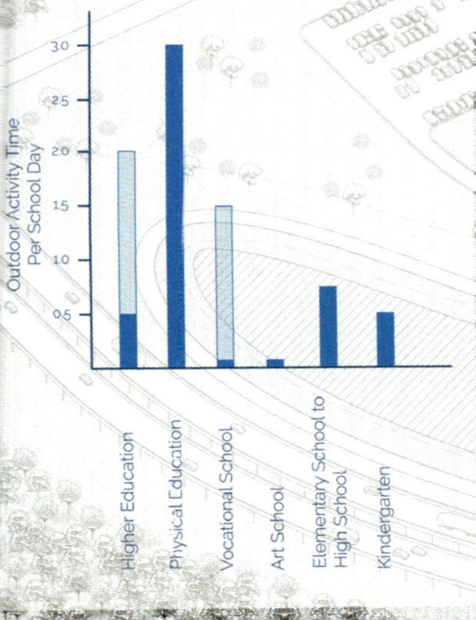

Education City

A Study of Korean Education Institutions and the Corresponding Outdoor Space

Following the advaneing need of education quality, Korean Education Institutions are at the steps of exploring the possibilities of campus design together with the revision of learning methodology. This study covers major categories of Korean Education Institutions ranging and how the campus facilities are designed to meet the academic need and at the same time, induce public involvement opportunities.

ziben

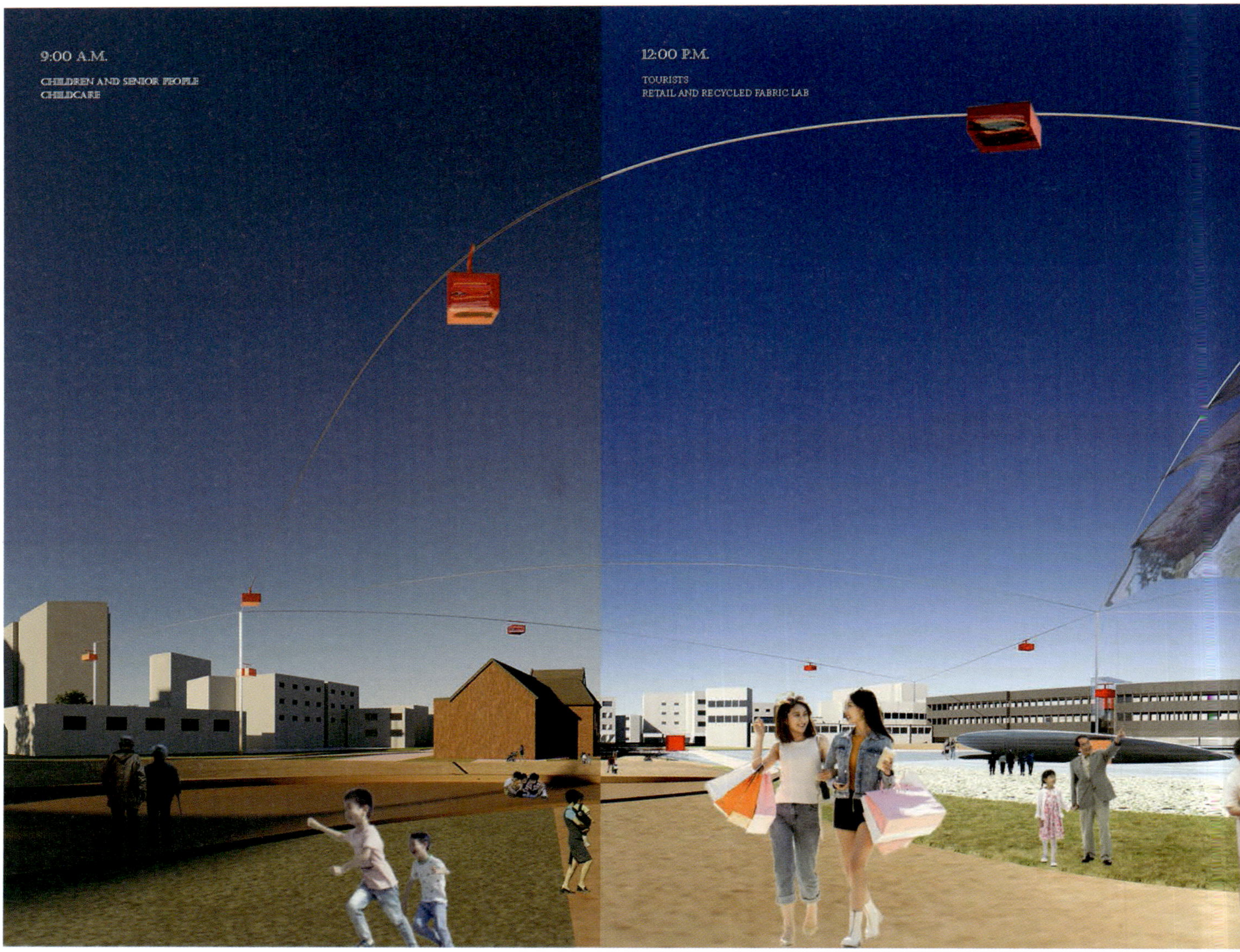

This project delves into a detailed analysis of a vibrant urban setting encompassing the fashion and food markets. We scrutinize user activities during both day and night, focusing on the diverse range of individuals populating this 24-hour hub, from tourists and citizens to fashion retailers and wholesalers.

Examining daily routines, we aim to capture the essence of this dynamic environment, where fashion designers, logistics workers, and more engage in their respective activities. Notably, amidst the hustle and bustle, there emerges a need for a relaxing space, prompting our exploration into reprogrammable projects globally.

Inspired by cases such as the German national parliament, Parking Day, Open Street programs, The Shed, and the Fun Palace, we categorize and scale these initiatives. The goal is to distill insights that contribute to the creation of a purposeful and adaptable space, emphasizing the enrichment of people's lives through a thoughtful integration of space and time.

Driven by the concept of leisure, our project is a living canvas adorned with recycled materials—repurposed wood, reclaimed metals, recycled plastics, and salvaged textiles.
Beyond sustainability, these elements infuse the project with a dynamic character that constantly evolves.

Experience the perpetual transformation of our project as recycled materials breathe new life into its essence.
Join us in weaving a sustainable future where creativity knows no bounds and materials yet to be invented find their place in our evolving narrative.

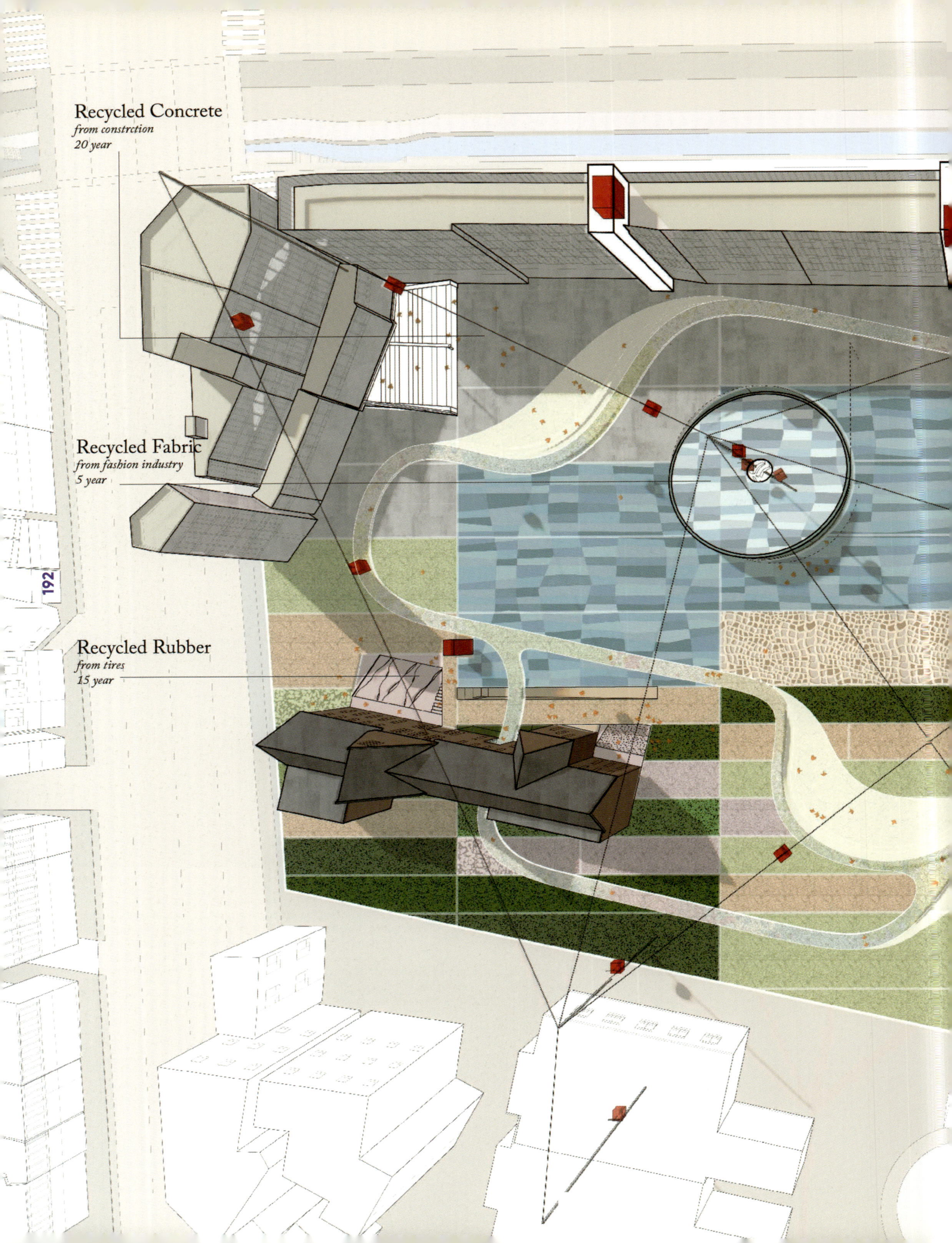
Recycled Concrete
from constrction
20 year
Recycled Fabric
from fashion industry
5 year
Recycled Rubber
from tires
15 year

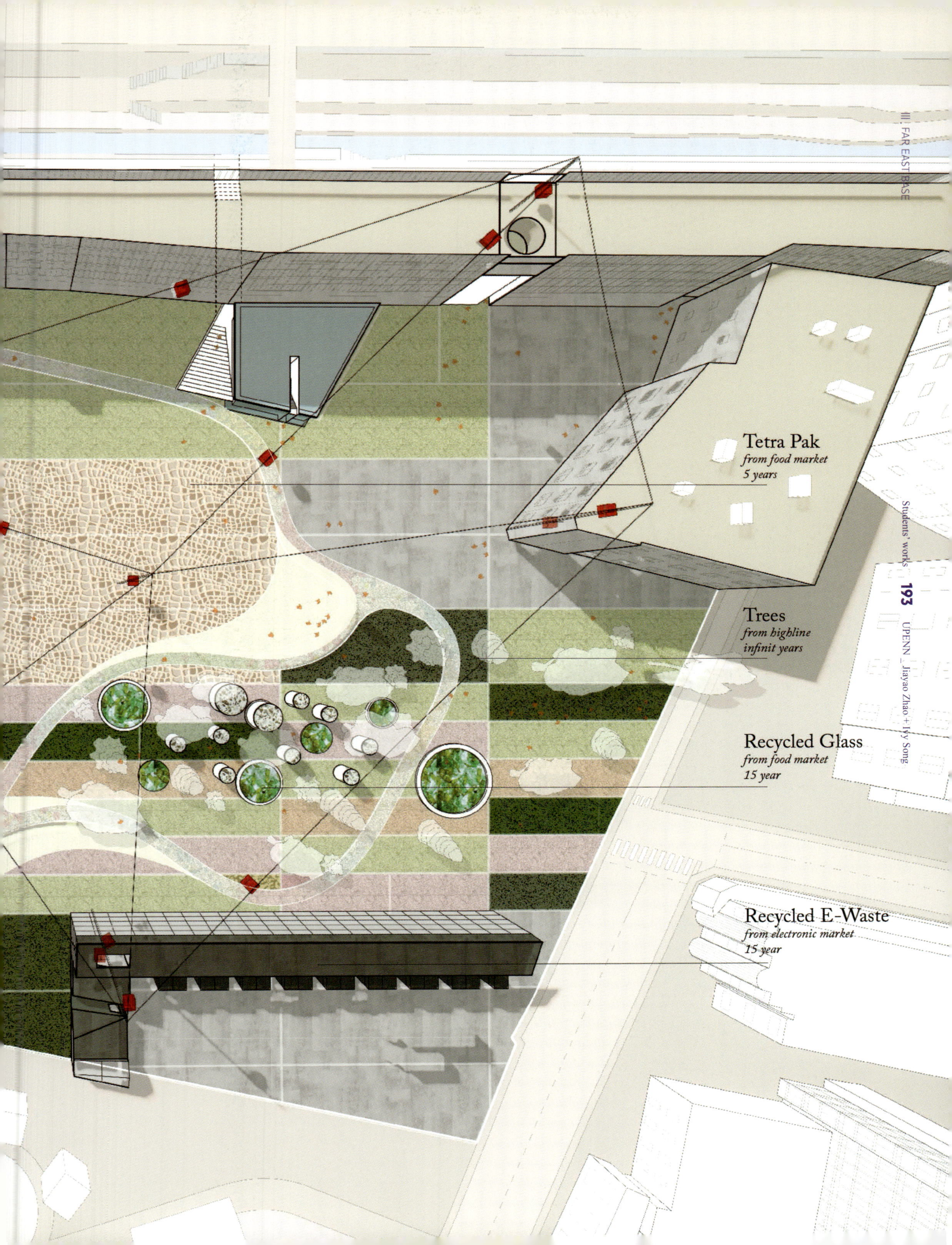
Tetra Pak
from food market
5 years
Trees
from highline
infinit years
Recycled Glass
from food market
15 year
Recycled E-Waste
from electronic market
15 year

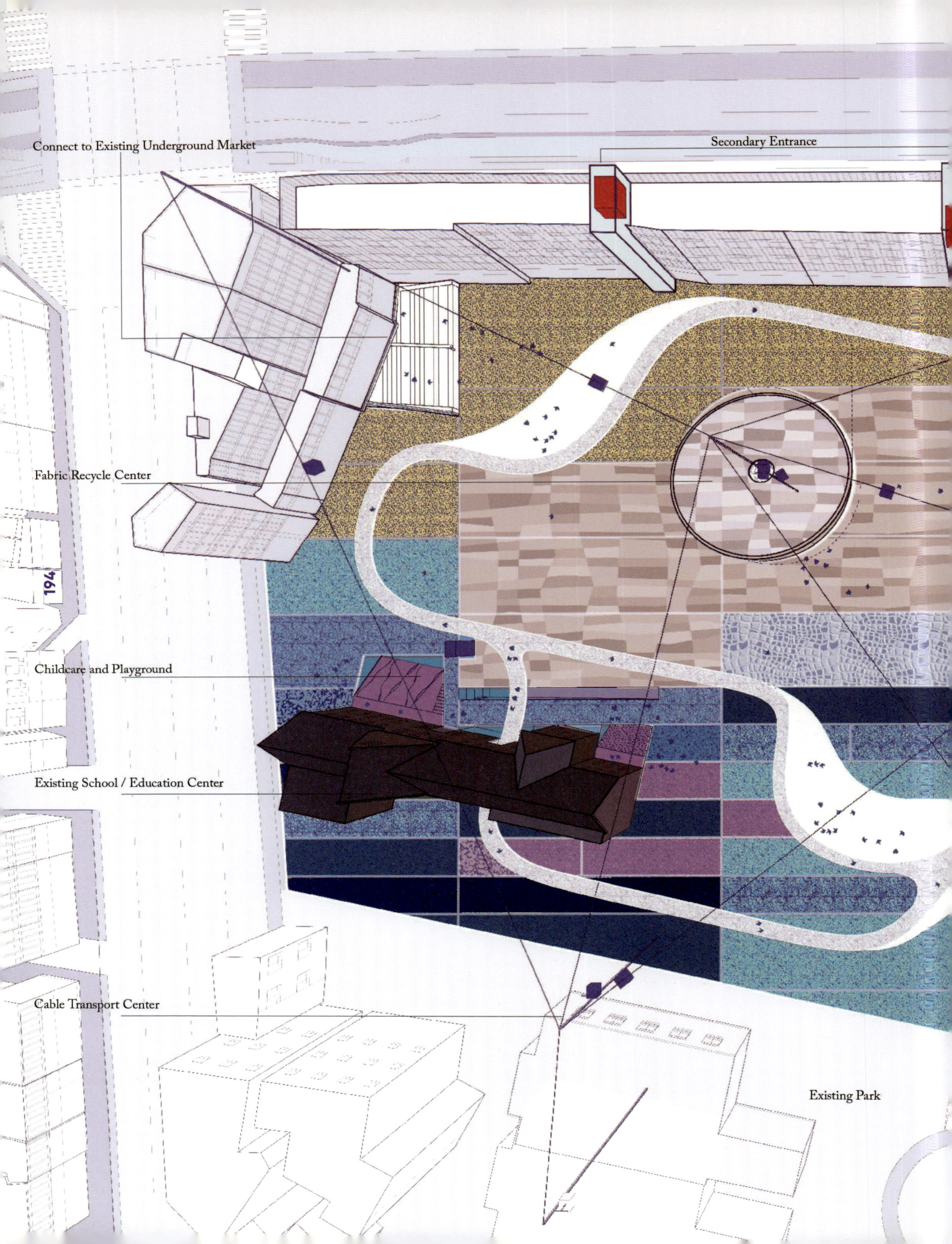
Connect to Existing Underground Market
Secondary Entrance
Fabric Recycle Center
Childcare and Playground
Existing School / Education Center
Cable Transport Center
Existing Park

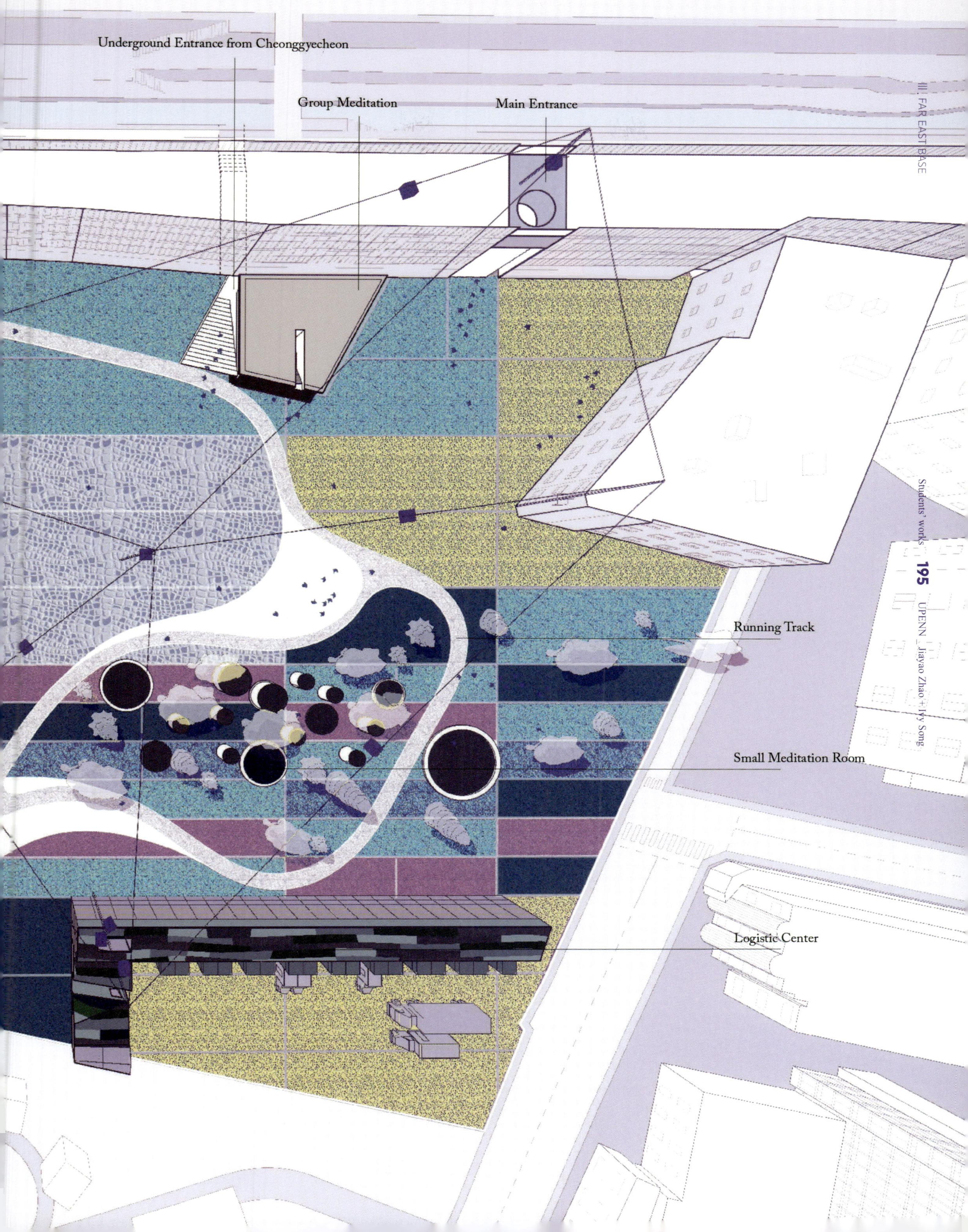
Underground Entrance from Cheonggyecheon
Group Meditation
Main Entrance
Running Track
Small Meditation Room
Logistic Center

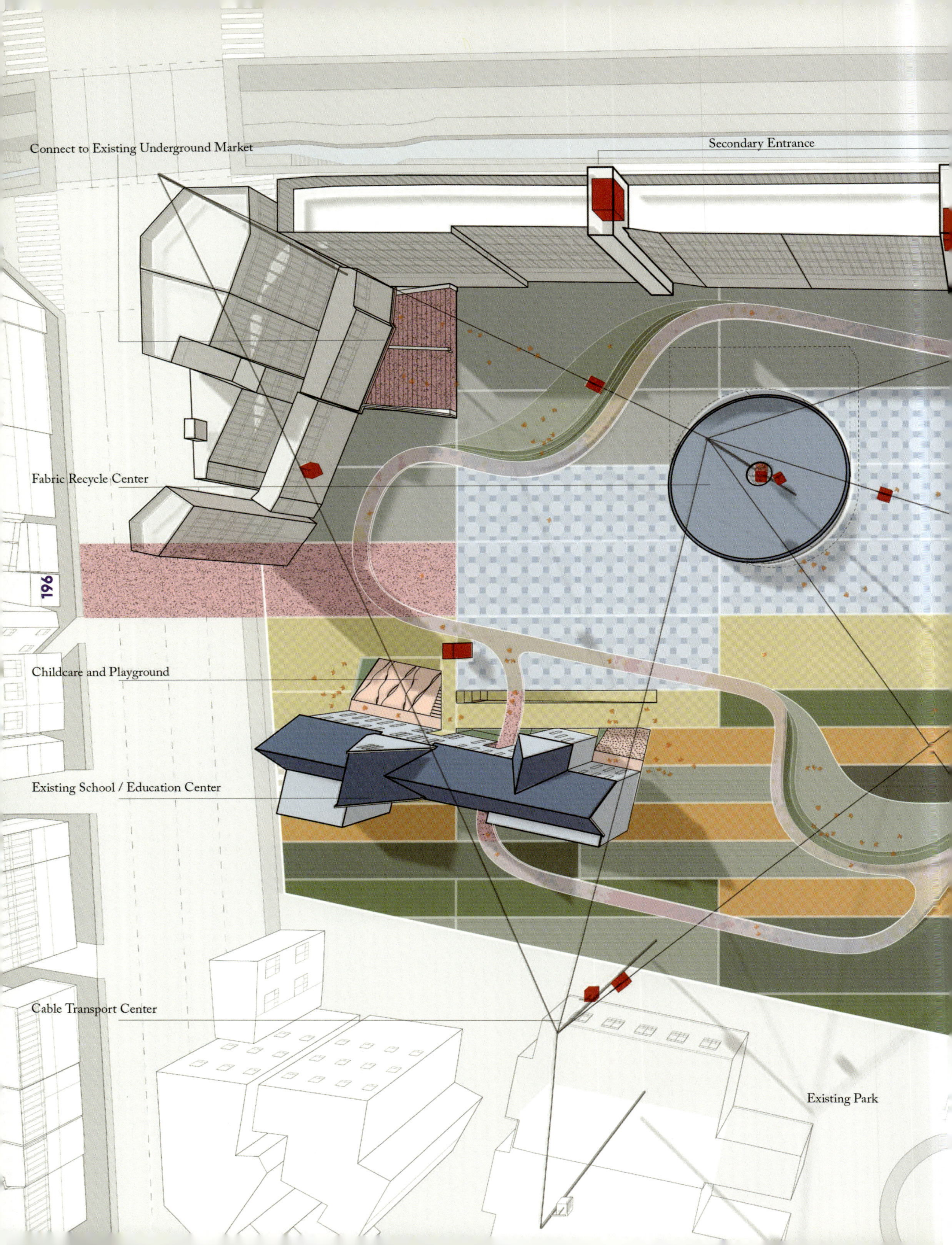
Connect to Existing Underground Market
Secondary Entrance
Fabric Recycle Center
Childcare and Playground
Existing School / Education Center
Cable Transport Center
Existing Park

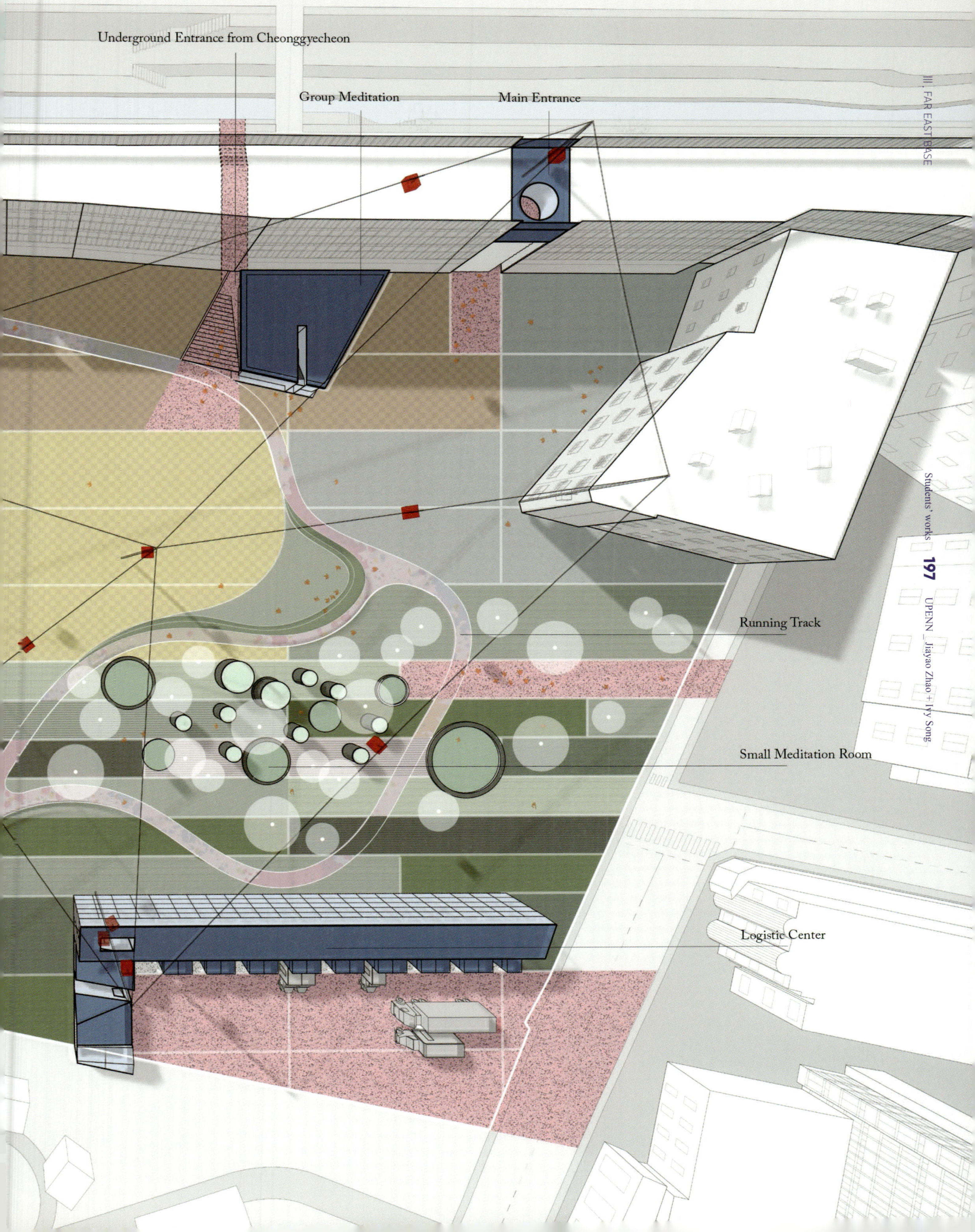
Underground Entrance from Cheonggyecheon
Group Meditation
Main Entrance
Running Track
Small Meditation Room
Logistic Center

Recycled Fabric
from fashion industry
5 year

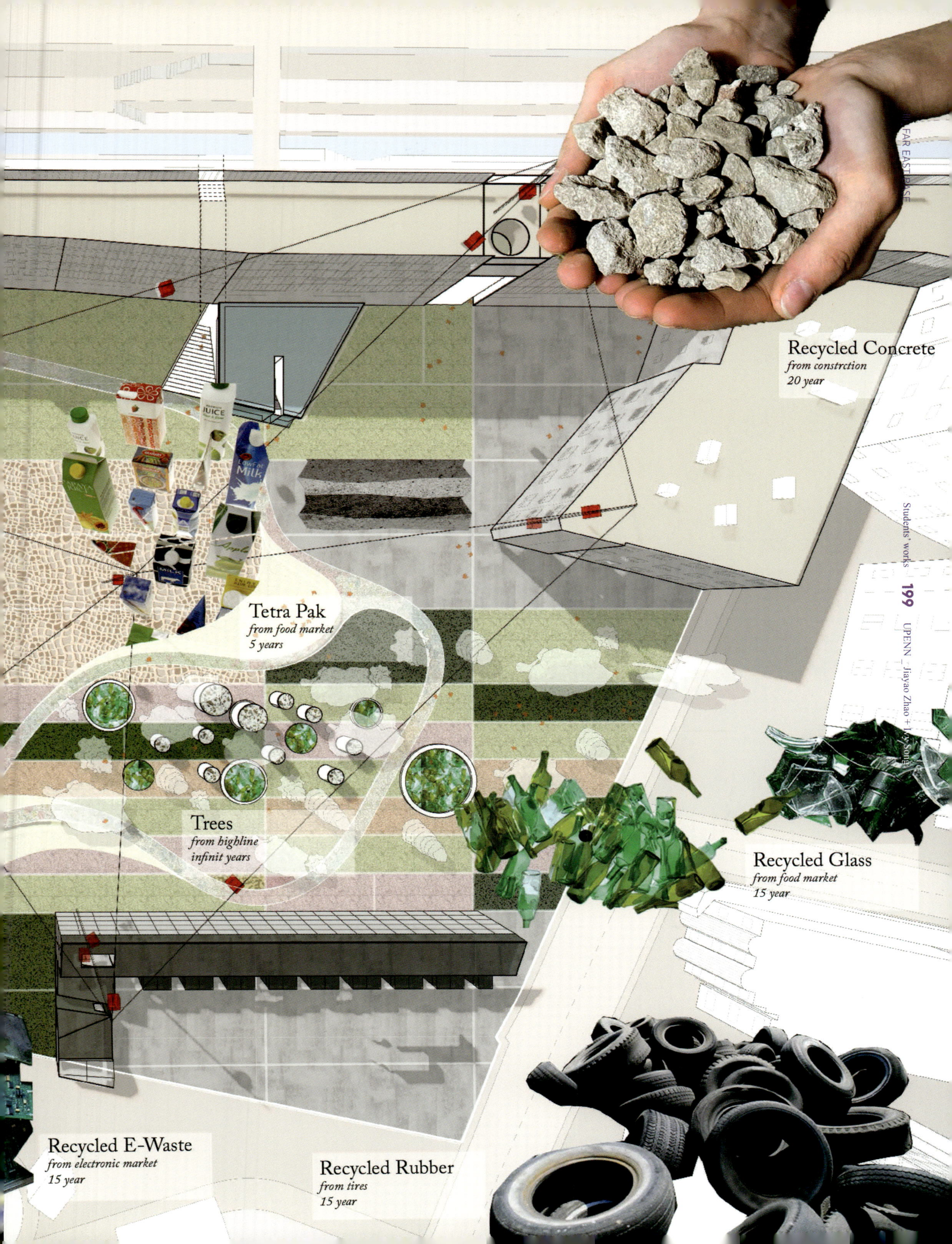
Recycled Concrete
from constrction
20 year
Tetra Pak
from food market
5 years
Trees
from highline
infinit years
Recycled Glass
from food market
15 year
Recycled E-Waste
from electronic market
15 year
Recycled Rubber
from tires
15 year

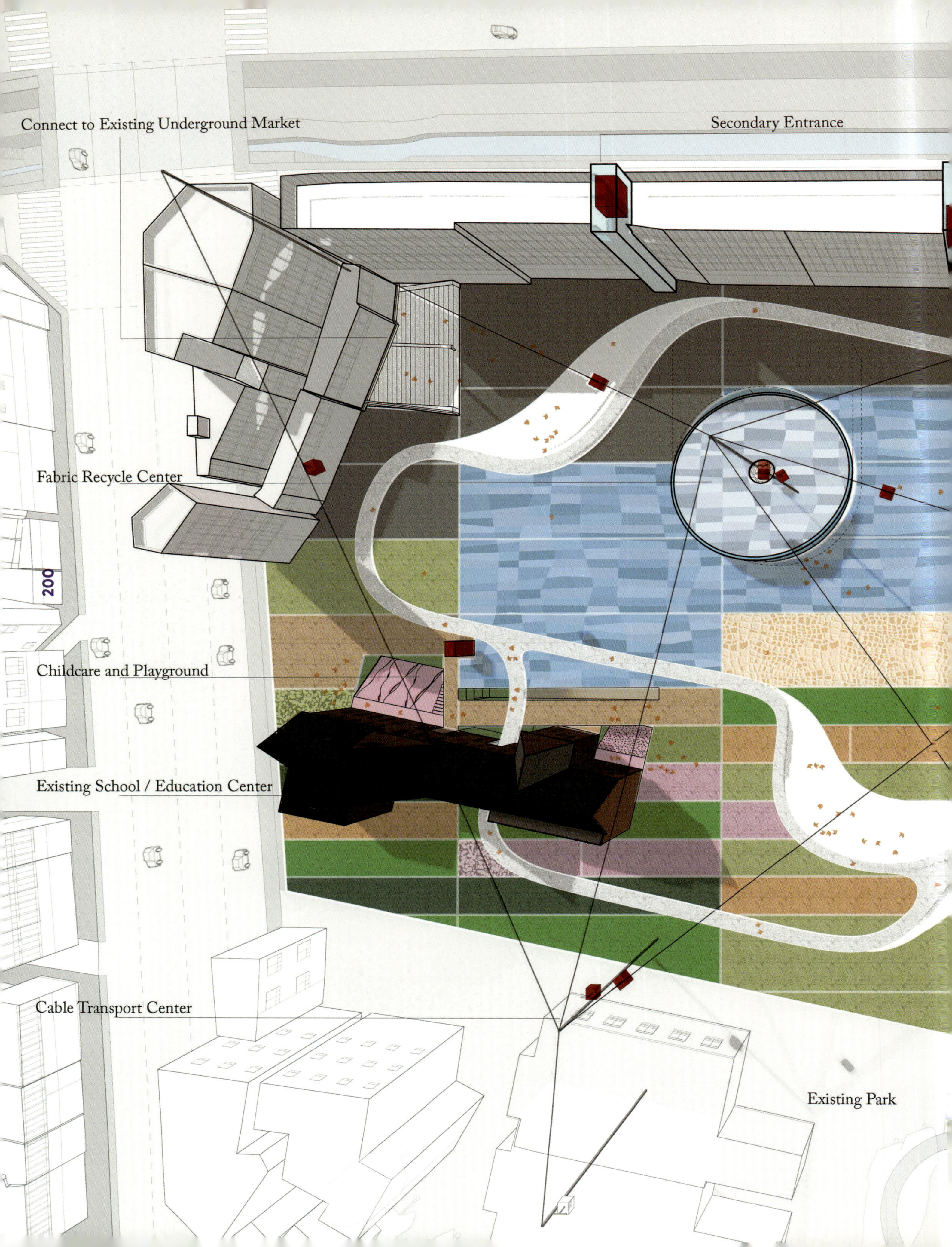
Connect to Existing Underground Market
Secondary Entrance
Fabric Recycle Center
Childcare and Playground
Existing School / Education Center
Cable Transport Center
Existing Park

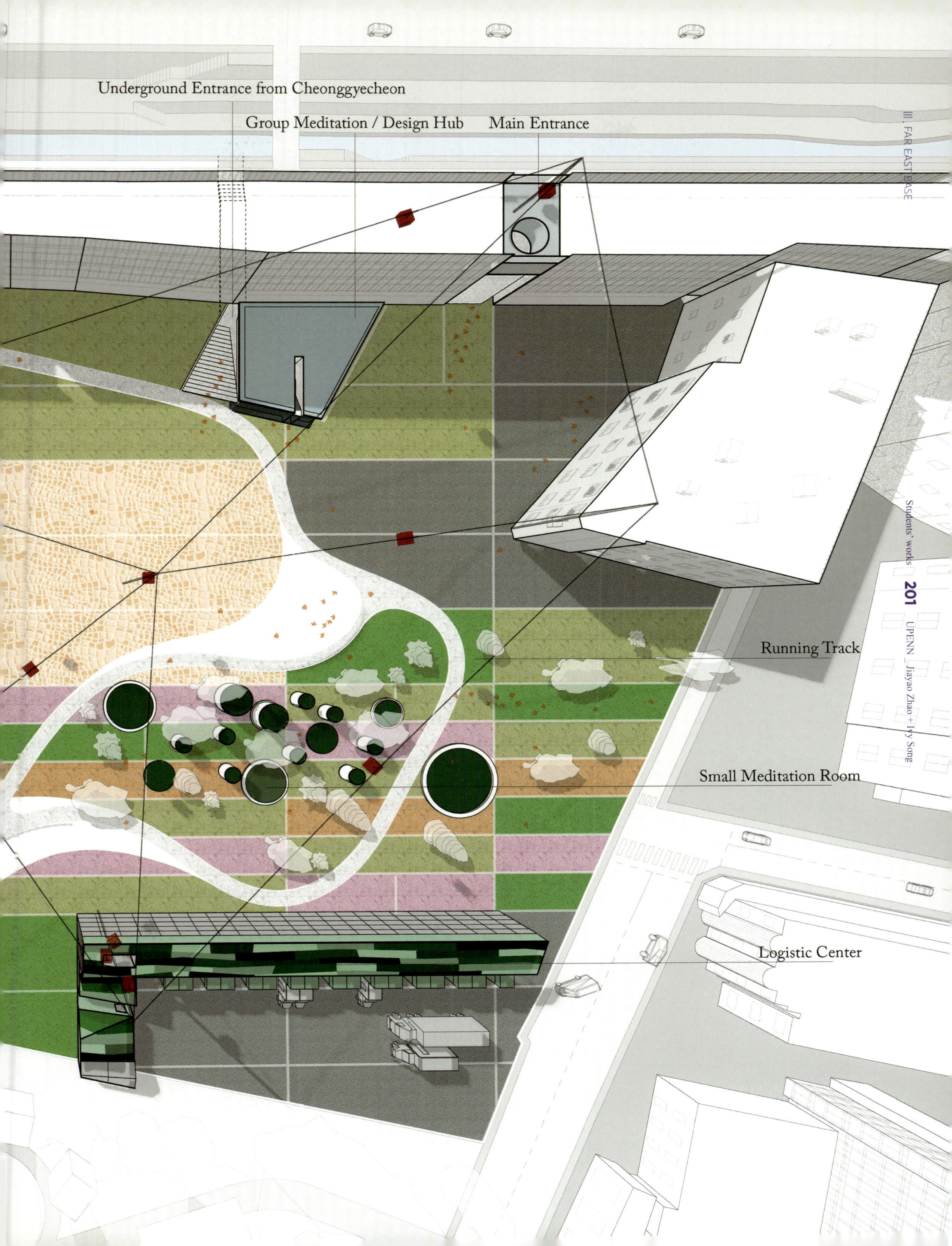
Underground Entrance from Cheonggyecheon
Group Meditation / Design Hub
Main Entrance
Running Track
Small Meditation Room
Logistic Center

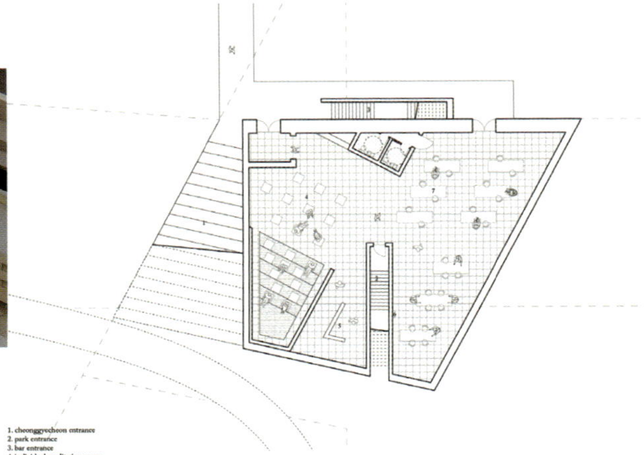

GROUP MEDITATION
DESIGN HUB

1. cheonggyecheon entrance
2. park entrance
3. bar entrance
4. individual meditation space
5. group meditation space
6. design pin up wall
7. group design space

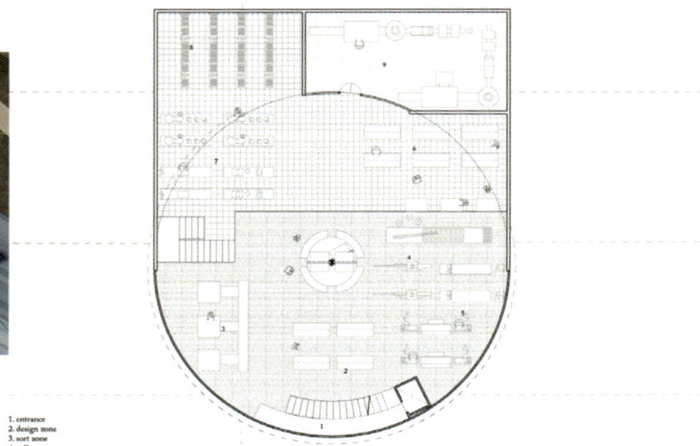

RECYCLED FABRIC LAB AND FACTORY

1. entrance
2. design zone
3. sort zone
4. collect zone
5. cut and press zone
6. preparing zone
7. filter, extract pulp, dry and iron zone
8. storage
9. textile process zone

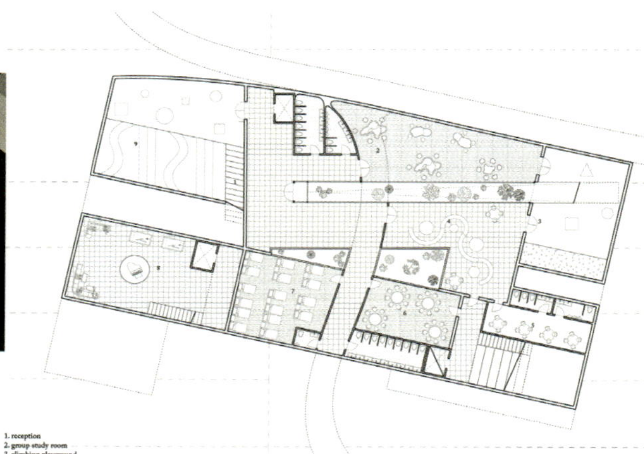

CHILDCARE

1. reception
2. group study room
3. climbing playground
4. library
5. discussion room
6. classroom
7. dorm
8. education center lounge
9. slide playground

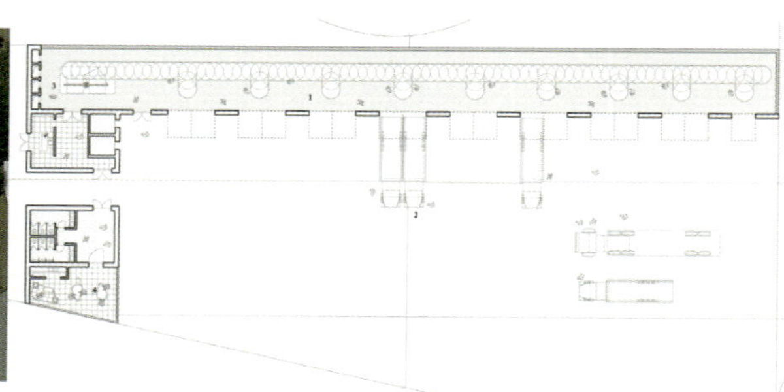

LOGISTIC CENTER

1. logistic transportation band
2. truck loading zone
3. distribution center
4. resting zone

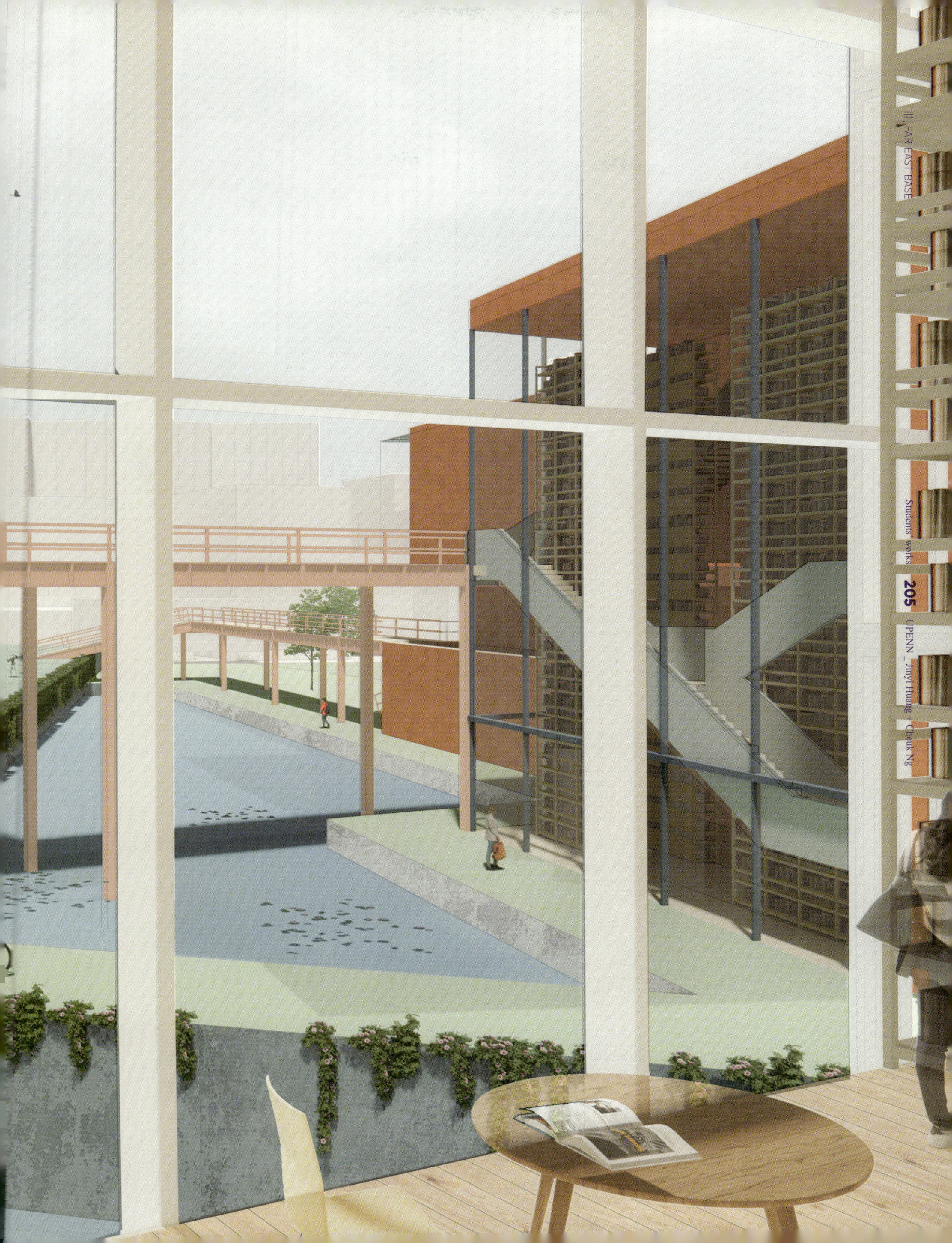

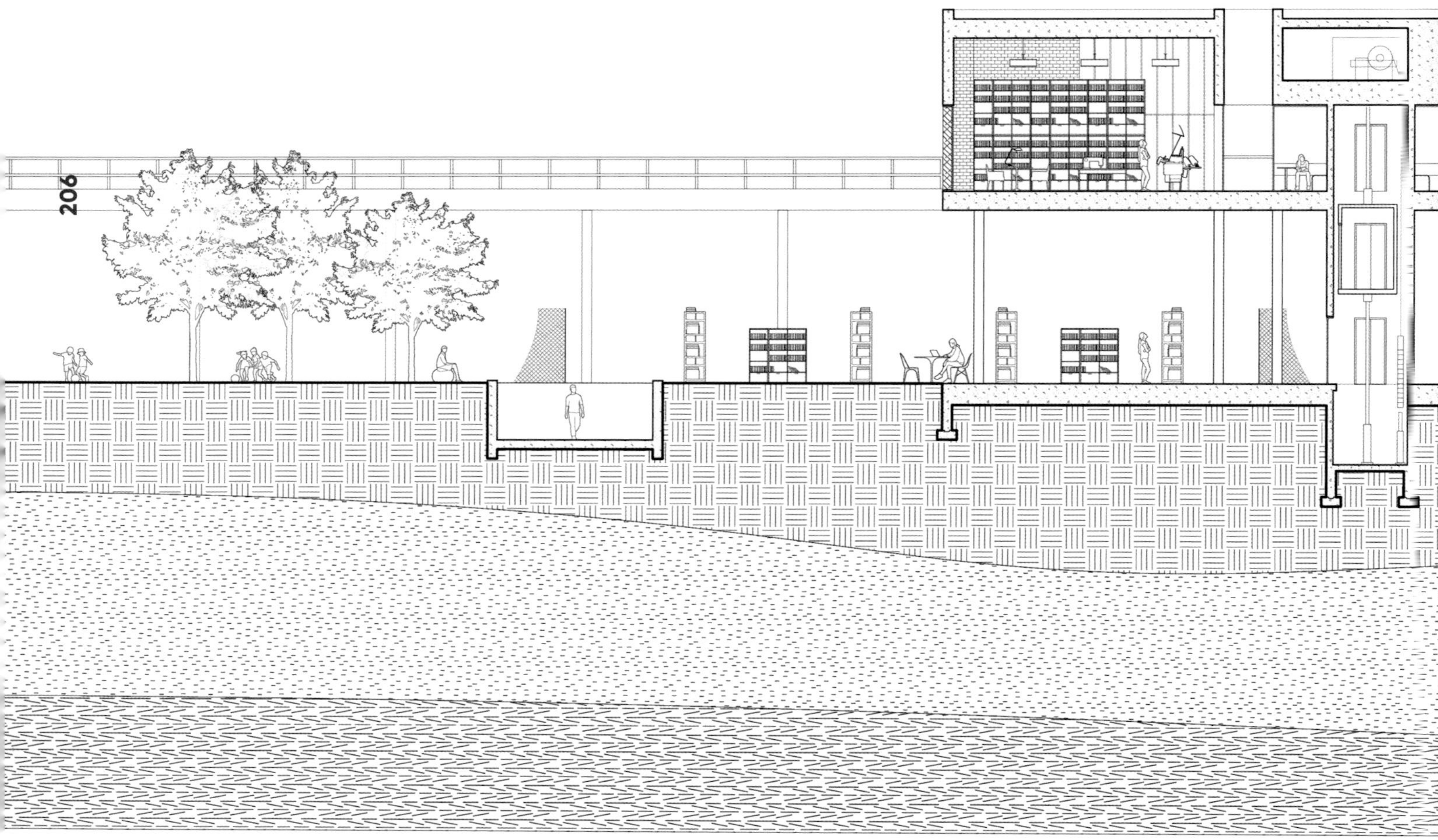

Economic disparities and education inequity is a pressing issue in Seoul.
Wealthier families can spend up to 5,000 dollars a month on cram schools, while the working class spends around 150 dollars. This unequal system leads to significant educational and social marginalization.

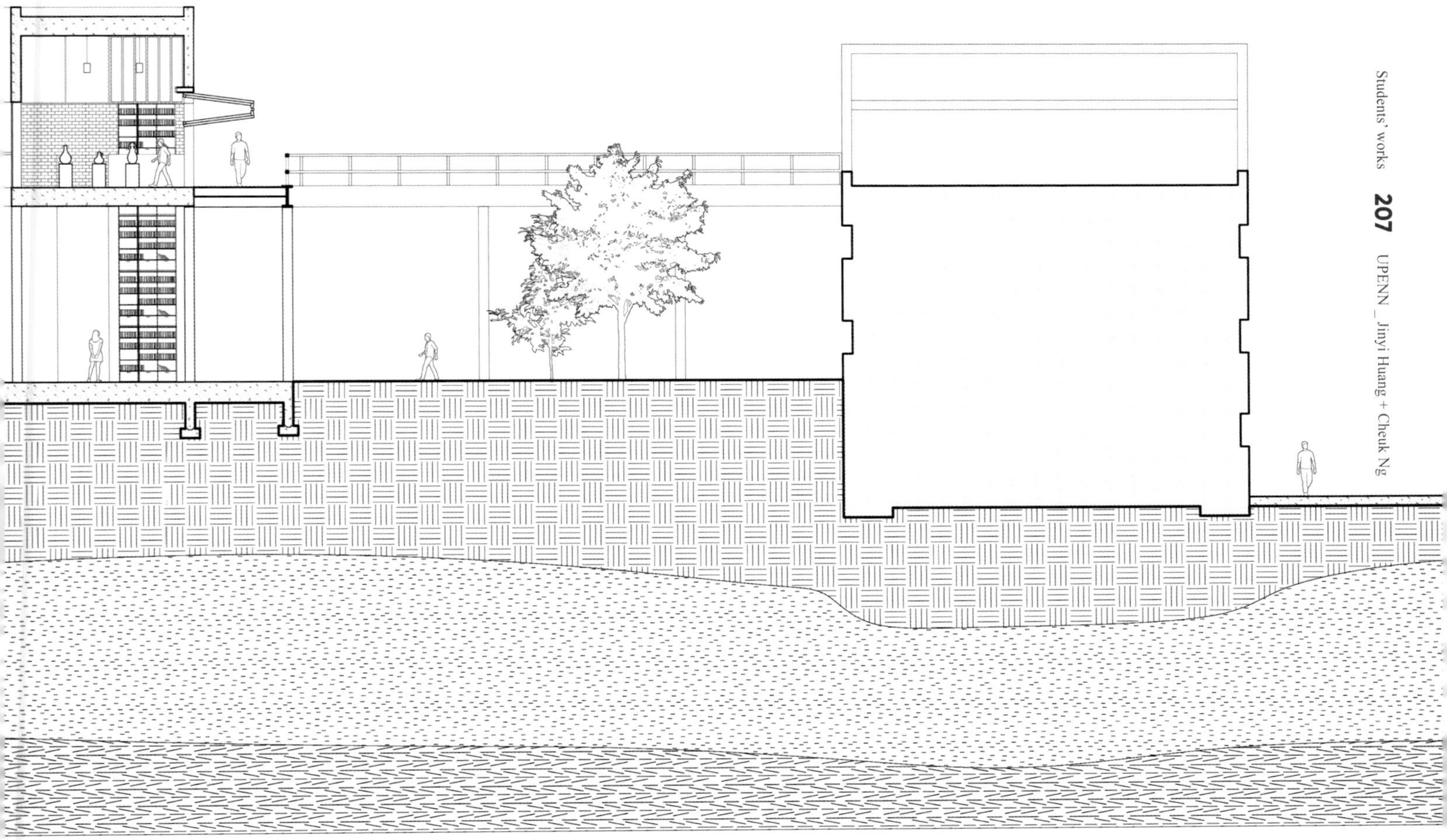

The site is located next to the Cheonggyecheon, an urban area known for its economic disparities and its controversial transformation from a slum to an expressway, and then to an artificial "riverwalk". Our project aims to create an equitable educational campus that addresses these disparities.
The space will accommodate various temporary events and can also be reprogrammed as a disaster relief shelter during the summer, when floods occur. Serving as the core value of the project, the reprogrammability and resilience of design are developed through providing infrastructural facilities that allow multiple kinds of events and education possibilities to be triggered.

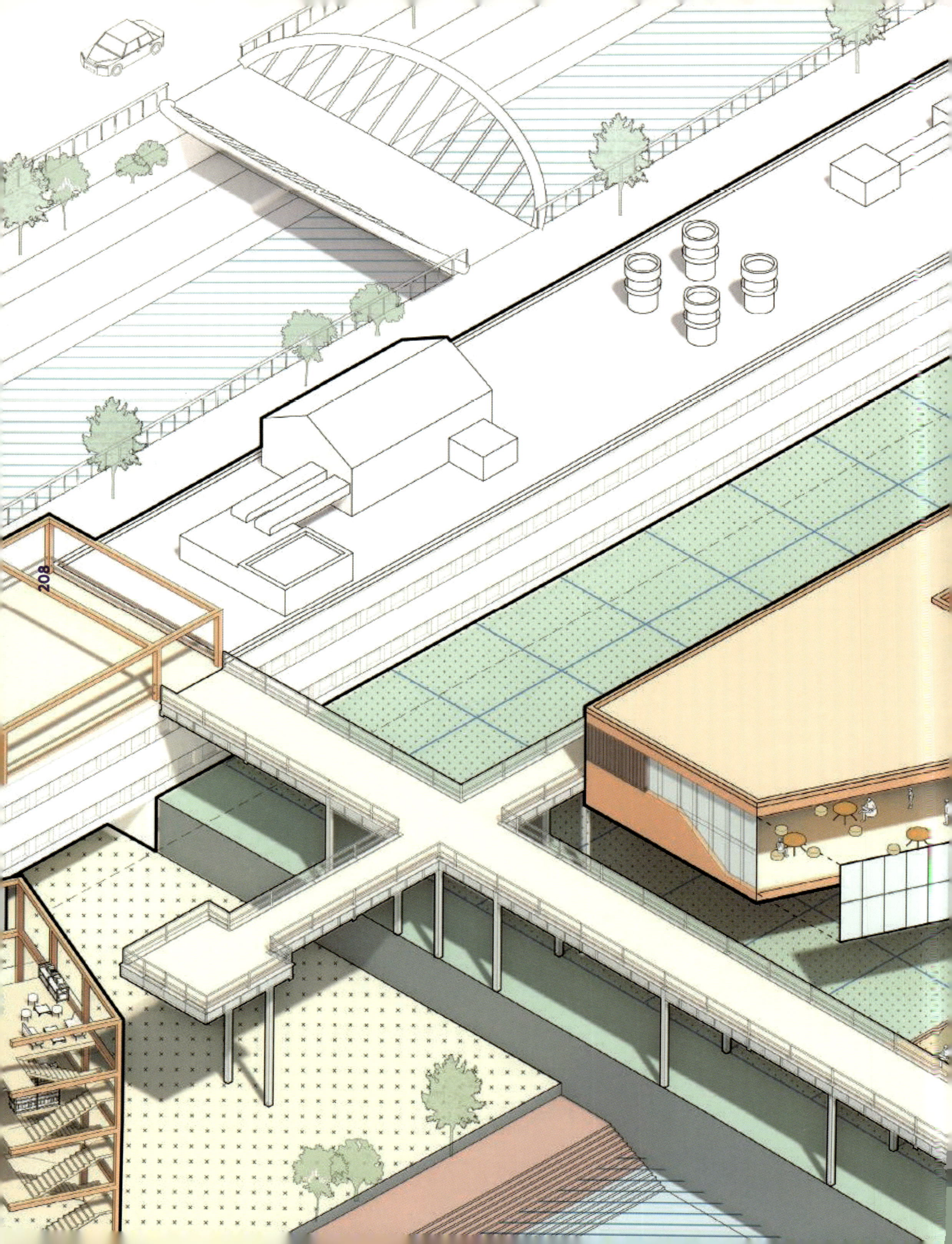

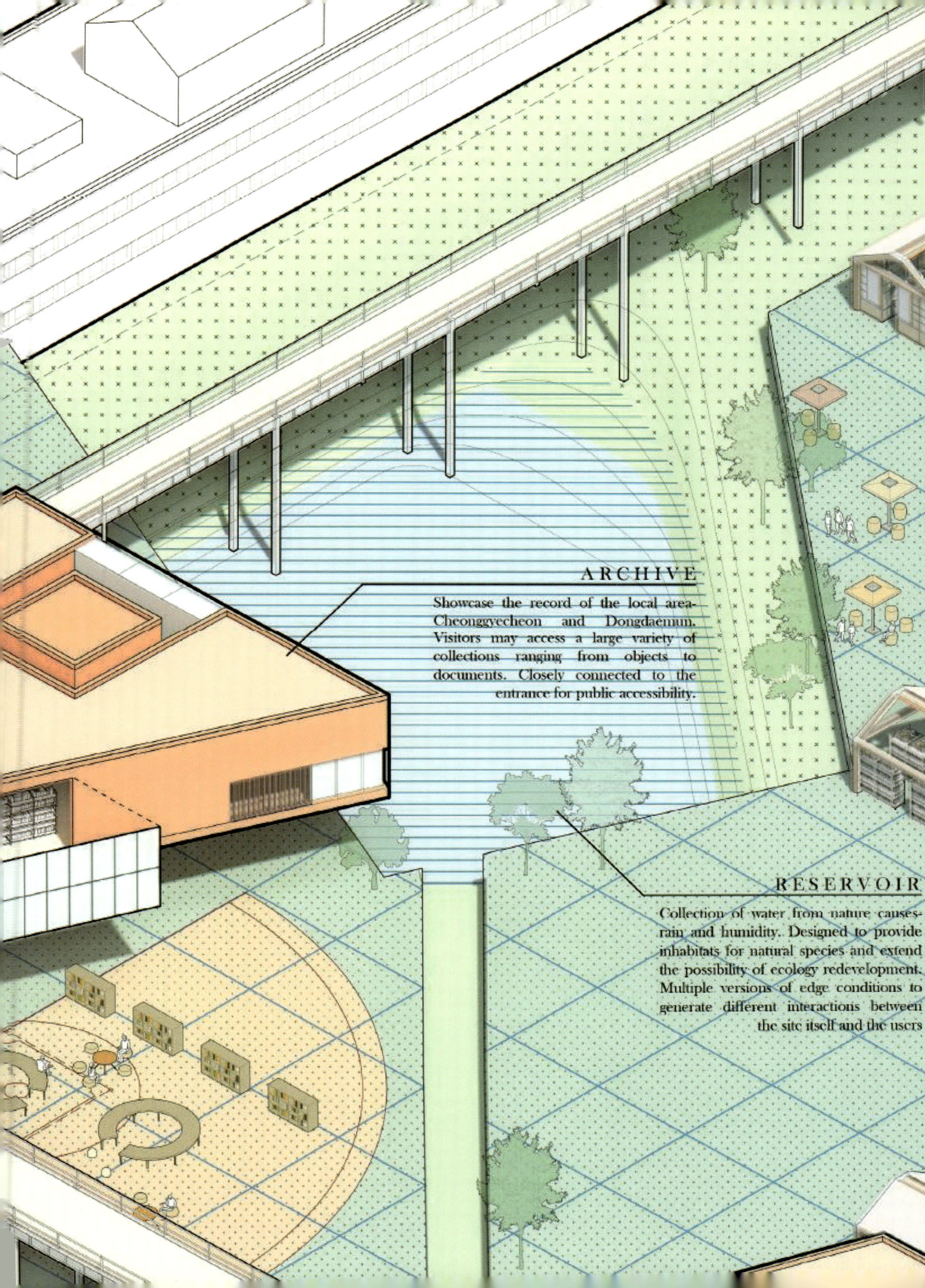
ARCHIVE
Showcase the record of the local area-Cheonggyecheon and Dongdaemun. Visitors may access a large variety of collections ranging from objects to documents. Closely connected to the entrance for public accessibility.
RESERVOIR
Collection of water from nature causes-rain and humidity. Designed to provide inhabitats for natural species and extend the possibility of ecology redevelopment. Multiple versions of edge conditions to generate different interactions between the site itself and the users

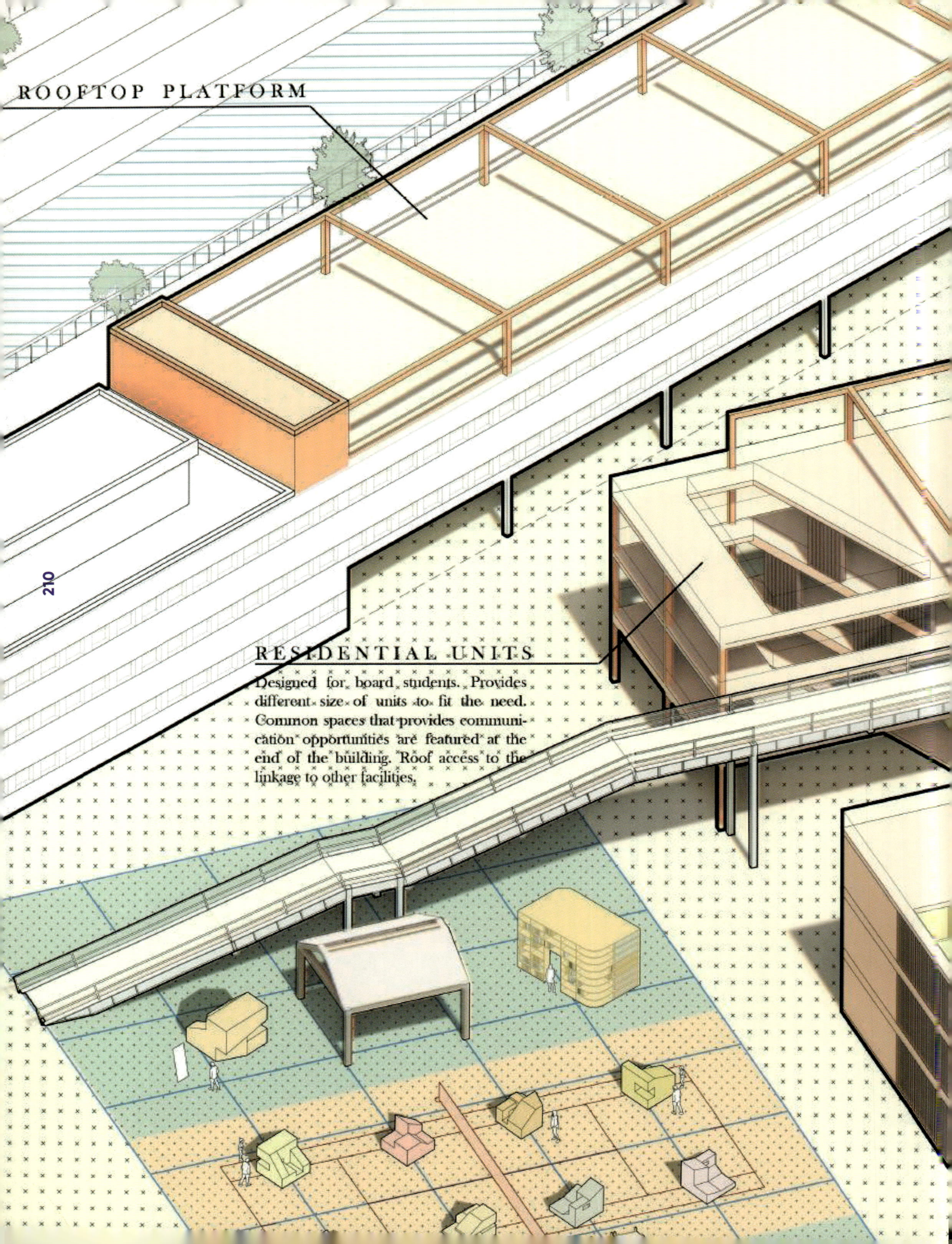
ROOFTOP PLATFORM
RESIDENTIAL UNITS
Designed for board students. Provides different size of units to fit the need. Common spaces that provides communication opportunities are featured at the end of the building. Roof access to the linkage to other facilities.

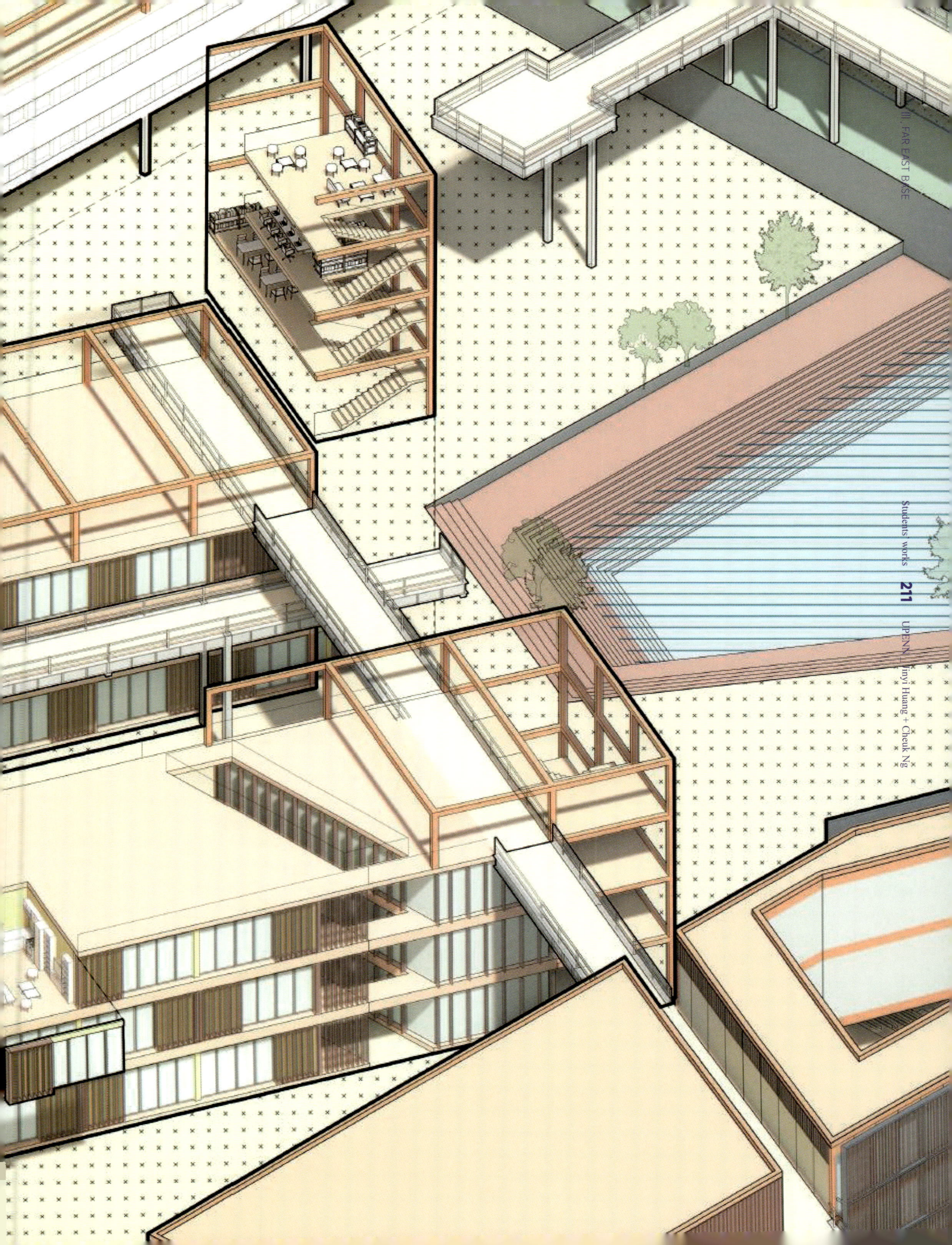

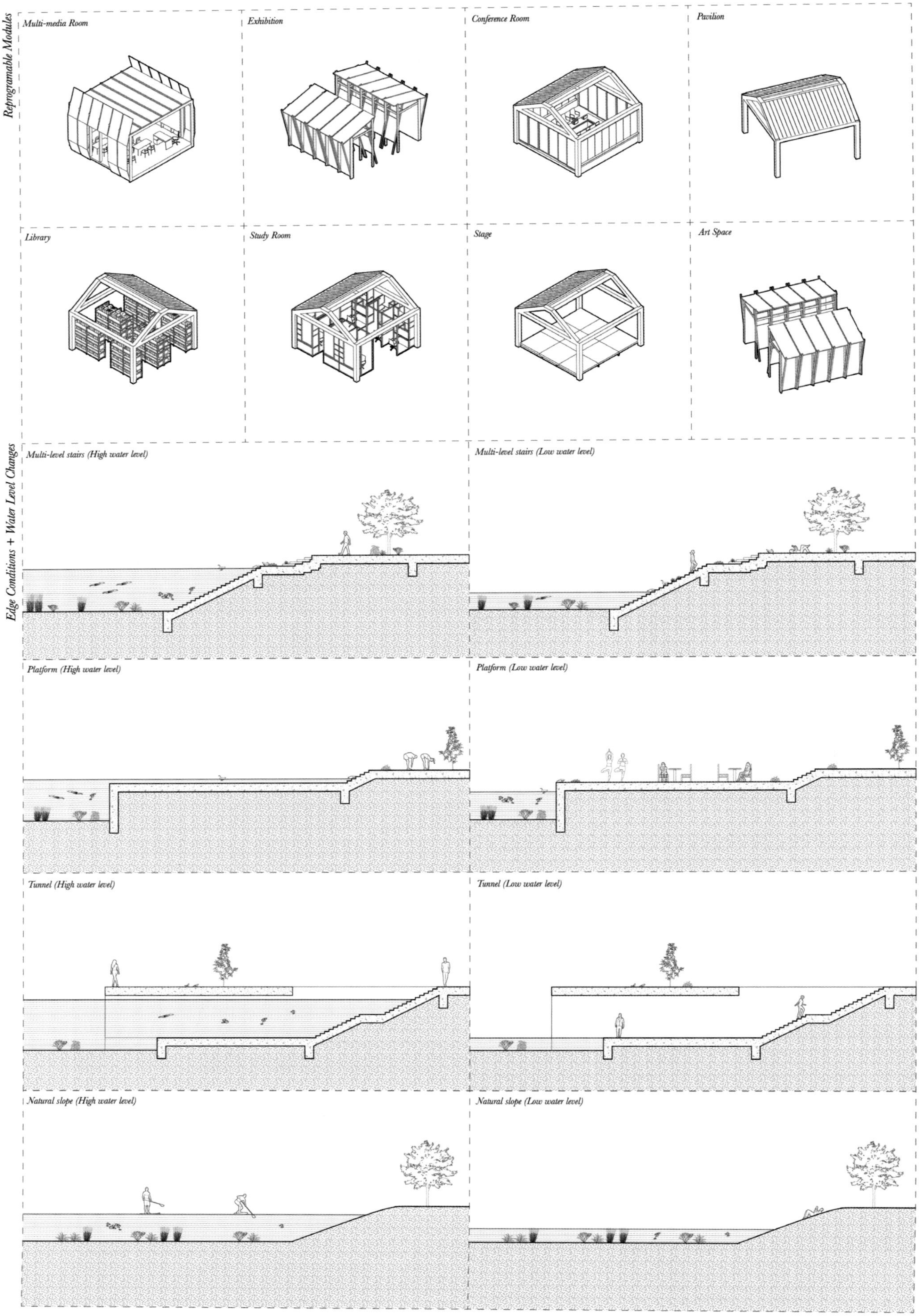
Reprogramable Modules
Multi-media Room
Exhibition
Conference Room
Pavilion
Library
Study Room
Stage
Art Space
Edge Conditions + Water Level Changes
Multi-level stairs (High water level)
Multi-level stairs (Low water level)
Platform (High water level)
Platform (Low water level)
Tunnel (High water level)
Tunnel (Low water level)
Natural slope (High water level)
Natural slope (Low water level)

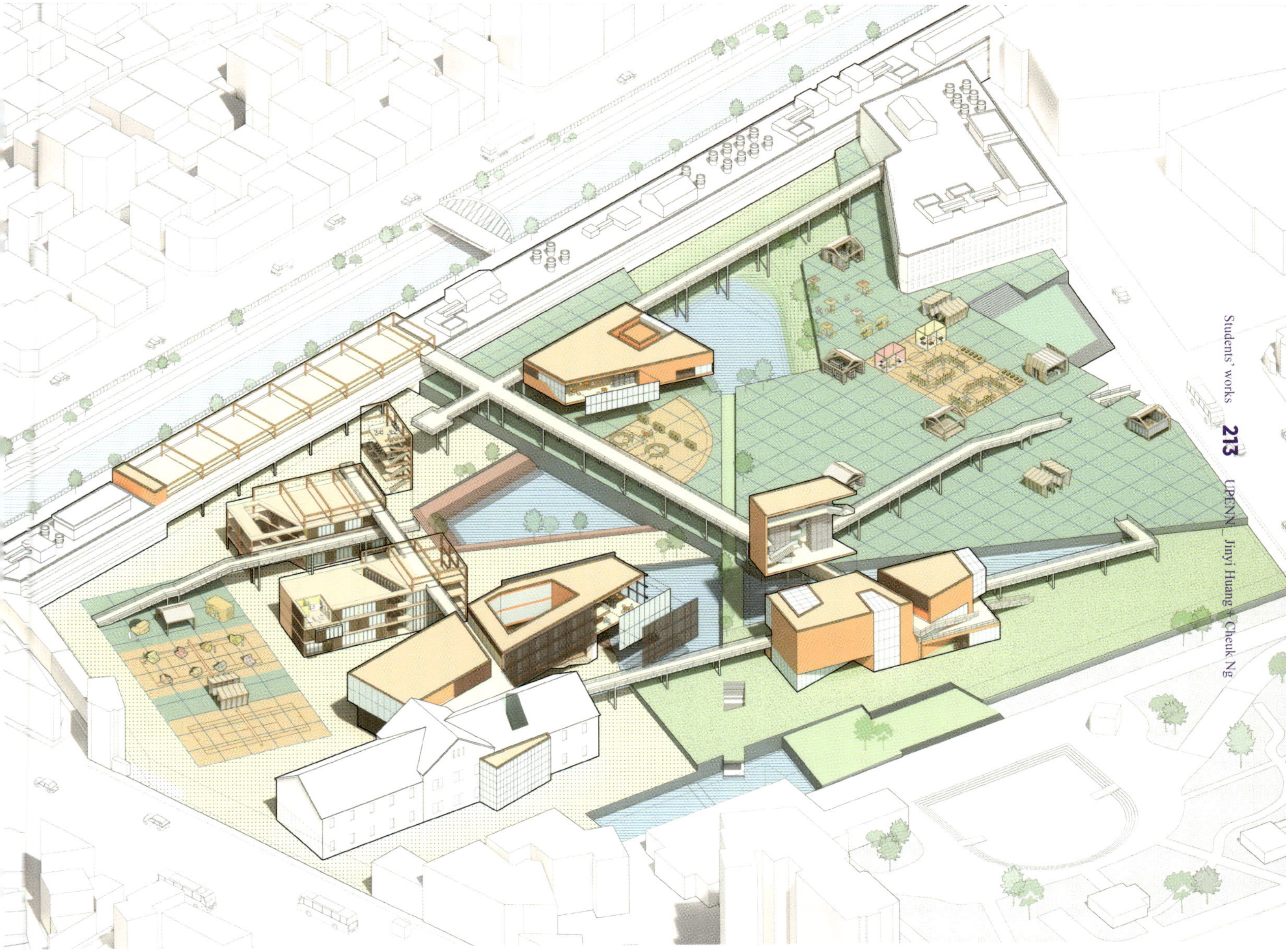

TERRA INCOGNITA

Seoul National University
DAAE
Yerin Kang, Associate Professor

STUDENTS

Juheon Kwon
Taeyang Lee
Shinwoo Park
Sangyup Yu
Youngjo Park
Donghyun Cho
Yeongchae Kim
Jeongwon Ko
Yoojin Chung
Jinho Kim
Yeojeong Heo(TA)

THE REPROGRAMMABLE CITY, SEOUL

University of Pennsylvania Weitzman
School of Design
David Eugin Moon, Visiting Professor

STUDENTS

Diego Martin
Hei Wai Valerie Tse
Maria Jose
Kelvin Vu
Aimen Arain
Reggie Zhao
Amy Greenberger
Pamela Pan
Jiayao Zhao
Ivy Song
Jinyi Huang
Cheuk Ng
Harsana Siva(TA)

ACKNOWLEDGEMENT

A sincere thanks to our sponsor Heerim Architects & Planners
and the help of Yongsan Park Management Office

SPONSOR

Heerim Architects & Planners Co., Ltd.
JEONG Young Kyoon, Chairman / CEO

EDITOR

Yerin Kang

In Seoul, there is a land that has existed all the time but has not been a space. It is not on a map. It is a military base largely left by the history of World War II and the Korean War, and the wall was built with the city for security reasons. Google Maps blocks spatial perception by visualizing the land as a dense forest of trees or bare land under construction. The lands that are not on this map are not connected with the surrounding urban groups but have been isolated like islands. However, with the succession of U.S. military bases returned, land like the island is gradually being opened to the city.

In this class, we are going to devise a way to make a land that is just open to the city, but an island-like land that is irrelevant to the context of the city, settle in the spatial order of Seoul. The studio aims to explore the possibilities of various types of urban projects beyond the development-oriented paradigm and to explore alternative architectural projects in which the structure of Seoul is supplemented and reshaped by the new land.

Students should come up with their own new Seoul Urbanism through research. Concepts such as "recovery" of the land, "connecting" with the surroundings, "building a relationship" with the city and nature, and emptying the land will be prioritized over "building and filling" in the land.

After deciding how to deal with the land through their own Seoul Urbanism, design an architecture or space that remembers the temporality of the land. What architectural strategies will be used to deal with memories and traces on a land where historical contexts are stacked? Human history is not alone in this place. The temporality of life forms other than humans that have lived on the land for a long time is also included. It should pay attention to the ecological map that has continued outside the history of humanism.

**THE REPROGRAMMABLE CITY, SEOUL-
TERRA INCOGNITA**

1 Gwanak-ro, Gwanak-gu
Seoul 08826
South Korea

Photo Credits
New york & Pennsylvania Trip
Students of Yerin Kang Studio 4-2 and David Eugin Moon Studio V

발행일 2024년 6월 20일 1쇄 발행
발행인 김기현
발행처 시공문화사 Spacetime Pub.
편집/디자인 아키스토리 jade-comm@hanmail.net

등록 1993년 3월 12일
주소 서울시 서대문구 독립문공원길 13 극동프라자 5층
전화 02) 3147-1212 **팩스** 02) 3147-2626

http//www.spacetime.co.kr
spacetime@korea.com

ISBN 978-89-5592-480-0
도서정가 20,000원